591.1

SOCIETY FOR EXPERIMENTAL BIOLOGY
SEMINAR SERIES · 5

ASPECTS OF ANIMAL MOVEMENT

ASPECTS OF
ANIMAL MOVEMENT

Edited by

H. Y. ELDER
Institute of Physiology
University of Glasgow

and

E. R. TRUEMAN
Beyer Professor of Zoology
University of Manchester

47,963

CAMBRIDGE UNIVERSITY PRESS

CAMBRIDGE
LONDON NEW YORK NEW ROCHELLE
MELBOURNE SYDNEY

Published by the Press Syndicate of the University of Cambridge
The Pitt Building, Trumpington Street, Cambridge CB2 1RP
32 East 57th Street, New York, NY 10022, USA
296 Beaconsfield Parade, Middle Park, Melbourne 3206, Australia

First published 1980

Set, printed and bound in Great Britain by
Fakenham Press Limited, Fakenham, Norfolk

British Library Cataloguing in Publication Data

Aspects of animal movement. – (Society for
Experimental Biology. Seminar series; 5).

1. Animal locomotion
I. Elder, H Y   II. Trueman, Edwin Royden
III. Series
591.1′852   QP301   79–41465

ISBN 0 521 23086 1 hard covers
ISBN 0 521 29795 8 paperback

Cilia
Amoeboid
Muscle

CONTENTS

CONTRIBUTORS

Alexander, R. McN.
Department of Pure and Applied Zoology, University of Leeds, Leeds
LS2 9JT.

Barlow, D. I.
Department of Biology, University of Southampton, Southampton SO9
3TU.

Bennet-Clark, H. C.
Department of Zoology, South Parks Road, Oxford OX1 3PS.

Currey, J. D.
Department of Biology, University of York, York YO1 5DD.

Dorsett, D. A.
Marine Biology Department, University College of North Wales, Bangor
LL57 2DG.

Elder, H. Y.
Institute of Physiology, University of Glasgow, Glasgow G12 8QQ.

Goldspink, G.
Muscle Research Laboratory, Department of Zoology, University of
Hull, Hull HU6 7RX.

Nachtigall, W.
Zoologisches Institut des Saarlands, 66 Saarbrucken, West Germany.

Rayner, J. M. V.
Department of Zoology, University of Bristol, Woodland Road, Bristol
BS8 1UG.

Sleigh, M. A.
Department of Biology, University of Southampton, Southampton SO9
3TU.

Taylor, C. R.
Museum of Comparative Zoology, Harvard University, Cambridge,
Massachusetts 02138, USA.

Trueman, E. R.
Department of Zoology, University of Manchester, Manchester M13 9PL.

Videler, J. J.
Department of Zoology, Groningen University, Haren, The Netherlands.

Wardle, C. S.
Marine Laboratory, PO Box 101, Aberdeen, Scotland.

PREFACE

This volume on 'Aspects of animal movement' represents the proceedings of a Seminar Series symposium held at the Reading meeting of the Society of Experimental Biology in December 1978. An additional paper on the aerodynamics of insect jumping, which was presented at a subsequent session, was also offered for publication and is included.

The papers given at the meeting represented the particular interests of contributors and their experimental results. As authors of the articles in this book they were invited to survey the areas of their individual interest and to give appropriate references in those fields for which space did not allow full coverage. We recognised that this treatment would leave gaps but at least this ensured that the principal advances would be adequately presented in the space of a short volume.

As editors we are grateful to all those who have given us advice and support, notably Professor R. McNeill Alexander and Dr J. W. Hannay the publications officer of the SEB. We also thank the authors for their prompt submission of manuscripts and their willing cooperation in making necessary amendments.

H. Y. Elder
E. R. Trueman

G. GOLDSPINK

Locomotion and the sliding filament mechanism

Introduction

The sliding filament theory of muscular contraction was proposed independently by A. F. Huxley & Niedergerke (1954) and by H. E. Huxley & Jean Hanson (1954). A description of muscular contraction may be found in most modern textbooks of physiology or cell biology and therefore it will not be described in detail here, however, the basic structure is shown in Figs. 1 and 2. Briefly, it involves sets of thick and thin filaments which slide over each other with the sliding motion being produced by cross-bridges. These cross-bridges which project from the thick filaments are believed to be independent force generators which either push or pull the thin filaments over the thick filaments (Fig. 3).

This basic mechanism seems to be the method of producing force in the contractile cells of almost all metazoan animals. As it has been so widely adopted throughout the animal kingdom, it may be assumed that this is an effective mechanism for producing force and also that it is a very adaptable mechanism. Certainly, when one considers the diverse types of locomotion, it is surprising that the same basic mechanism has been adapted to so many different situations. In this paper, I will review some of the unique features of the sliding filament mechanism and attempt to show how these are modified for different locomotory requirements.

Maximum force and overall rate of shortening

The maximum force production of a muscle is directly proportional to the number of cross-bridges in parallel and is of the order of $500\,kN\,m^{-2}$

cross-sectional area of muscle tissue. Ways of arranging muscle fibres so that they have a larger effective cross-sectional area and hence more cross-bridges in parallel have been discussed elsewhere. (Alexander, 1974; Calow & Alexander, 1973.)

The rate of shortening of a muscle is perhaps more interesting because the range found throughout the animal kingdom is very wide. At one extreme, we have the very rapid contraction of mouse limb muscles or insect flight muscles which take only a few milliseconds to contract and at the other extreme we have the very slow contraction of the chicken anterior latissimus dorsi or the limb muscles of the tortoise which may take several seconds to complete their contraction.

The rate of shortening of a muscle depends on several factors. These include the number of sarcomeres in series, the length of the filaments of the sarcomere and the rate of shortening of each new sarcomere. This latter

Fig. 1. Diagram of the ultra structure of striated muscle showing the arrangement of the thick and thin filaments into sarcomeres. The relationship of the sarcoplasmic reticulum to the myofibrils is also shown. From Peachy (1965).

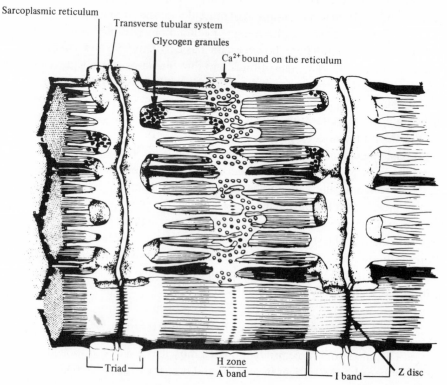

aspect is referred to as the intrinsic rate of contraction and it is presumably determined by the rate at which the cross-bridges develop force and the duration of the cross-bridge cycle. During each cycle, each cross-bridge hydrolyses one molecule of adenosine triphosphate (ATP) and therefore the intrinsic rate of contraction may be estimated by measuring the specific activity of its myosin (myofibrillar) ATPase. The physiological method of measuring the intrinsic rate usually involves measuring the rate of shortening in the muscle with different loads and then extrapolating back to zero load. The rate of shortening is then expressed in either muscle lengths per second or sarcomere lengths per second. The molecular structures involved in determining the rate of working of the cross-bridge are not fully understood but it is known that there are some small proteins (light chains) associated with the myosin molecule and that these differ in different

Fig. 2. Diagram of the molecular arrangement of the thick and thin filaments. The thick filaments are made up of myosin molecules which are shown as having a double helical structure, part of which is the backbone (light meromyosin). This is embedded in the thick filament which is made up of myosin backbones arranged in a helical way. The remaining part constitutes the cross-bridge (heavy meromyosin). This has a hinge part which is also compliant and two heads (S1 fragments) which are the active sites. The thin filament consists of a double helix of actin monomers. There are however proteins associated with the filaments; some of these regulate the 'switching on' and 'switching off' of contraction, others control the rate at which the contractile process proceeds. The two main sets of proteins are the troponins (Tni, TnC, TnT) and tropomyosin which in mammalian muscle regulate the switching on and switching off of contraction and the light chains of myosin which appear to be involved in controlling the speed of contraction, although in some cases the roles of these proteins have been changed during the course of evolution.

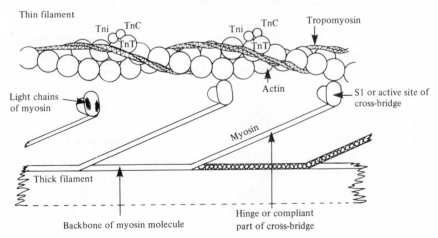

muscles. Cardiac muscle and most slow contracting muscles possess only two light chains, whilst fast contracting muscles have three light chains. The rate at which muscles use ATP is of course very important, as it determines, in energy terms, how expensive a contraction will be. As we shall discuss later, this is one of the main considerations in designing muscles for different locomotory and postural activities.

The question of filament lengths and sarcomere number is an interesting one. The sarcomeres in most animals are of a standard size, that is to say, the thick filaments are approximately 1.55 μm in length and the thin filaments are approximately 1.1 μm in length. In some crustacean muscle fibres the myosin filaments are considerably longer than 1.55 μm (Jahromi & Atwood, 1969; Warner & Jones, 1976). This means that there are more cross-bridges per sarcomere and hence each sarcomere is able to produce a greater amount of force. However, longer sarcomeres mean that there will be fewer of them along the length of the fibres and this results in a lower overall rate of shortening. For most muscles with standard filament lengths the overall rate of shortening is primarily determined by the intrinsic rate of shortening and the number of sarcomeres in the series.

Extent of shortening of different types of muscle

The extent to which a muscle can shorten is important but in addition the force developed over that range of shortening must also be considered. Striated muscle with standard filament lengths is able to shorten by 60%. The limit of the shortening is when the thick filaments butt against the Z discs (Fig. 3). In some muscles, for example those of the barnacle, a much greater degree of shortening is apparently possible because the disc lattice opens up and allows the thick filaments to pass through (Hoyle & Smyth, 1963; Hoyle, McAlear & Selverston, 1965).

Smooth muscle is also capable of a considerable degree of shortening and it also has the ability to develop reasonable force over almost the full range of lengths. The reason for this seems to be that an actin filament after interacting with one myosin filament may go on to interact with the next myosin filament. Helical or obliquely striated muscle also seems to be able to shorten to a much greater extent than striated muscle. The reason for this is that some of the length change results from a change in the pitch of the helix (Fig. 4). Both smooth muscle and helical muscle are used for locomotion in the invertebrates, particularly in soft-bodied animals where the hydrostatic mechanisms involve very considerable changes in the shape of the animal.

At the other extreme we have insect 'fibrillar' flight muscle in which the force developed shows a very sharp dependence on initial length (Fig. 5). This is a highly specialised type of muscle in which the sarcomeres shorten by only a small amount in order to move the wings up and down. In indirect flight muscles, the muscles produce a distortion of the thoracic skeleton which causes the wings to click up and down. The sarcomeres 'vibrate' at a high frequency and the contraction cycles are not synchronous with the nerve action potentials. After the initial activation by the nervous system further activation is instead believed to be caused by the wing lever system stretching the sarcomeres. Insect flight muscle is in fact an excellent example of how the sliding filament mechanism has been adapted to carry out a highly specialised mode of locomotion. Good reviews of this type of muscle may be found in Pringle (1972, 1975a, b).

Power output of muscles

Another important physiological parameter is power output. This involves both the force developed by the muscle and its rate of shortening. The more rapidly a muscle contracts and the more forcefully it contracts, the

Fig. 3. Shows the relationship between the thick and thin filaments in a relaxed muscle, contracted muscle and a fully contracted muscle. In most muscles the limit of shortening is reached when the myosin filaments butt against the Z discs.

Striated muscle

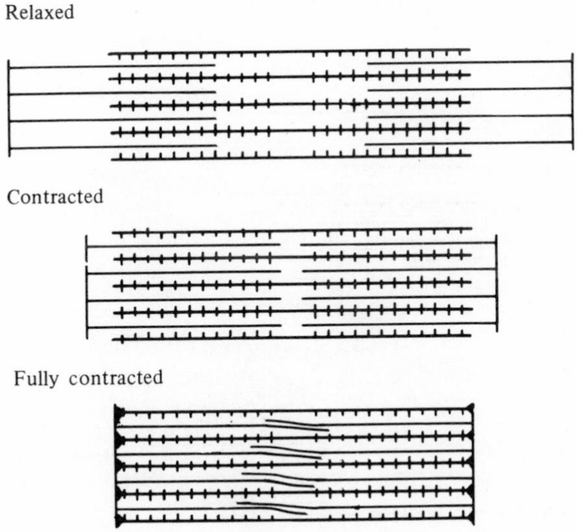

Relaxed

Contracted

Fully contracted

greater its power output. The maximum sort of power developed by muscles seems to be of the order of $1 \, W \, g^{-1}$. However, the maximum sustainable power is probably of the order of $0.3 \, W \, g^{-1}$ and this is probably only achieved in a few instances such as the flight muscles of the humming bird. The limiting factor as far as sustained power output is concerned is the supply of oxygen and metabolites rather than any deficiency in the sliding filament mechanism. Power output is not always the most important characteristic of a muscle. In some cases rapid contractions are not required and the efficiency of converting energy into work becomes the more important factor. It is for this reason we find that some muscle fibres are specialised for a high-power output whilst others are specialised for producing slow but very efficient contractions. This division of labour between muscles and muscle fibres is discussed below.

Fig. 4. Shows the arrangement of filaments in invertebrate helical and smooth muscle. In helical muscle some of the shortening is accounted for by the change in the pitch of the helix. In smooth muscle the thin filaments can pass from one thick filament to the next hence a reasonable amount of force can be produced over a wide range of shortening.

Helical muscle

Relaxed

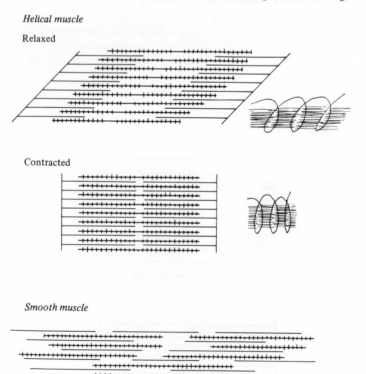

Contracted

Smooth muscle

Gradation of contraction

The factors which determine the maximum force a muscle can develop have been discussed. However, muscles are not often required to develop their maximum force and there must be some way of matching the force developed to that required for any particular activity.

In many invertebrate muscles this involves activating the muscle fibres to different extents. Invertebrate muscle fibres have usually more than one motoneuron. The properties of the muscle fibre membrane are such that the potentials generated at an end plate do not spread very far if only one or two nerve impulses are received. This is why these are sometimes referred to as non-twitch muscles. However, as the frequency is increased, more and more

Fig. 5. Length/active tension curves for a range of different muscles. The tensions are not given as grams per unit cross-section therefore they cannot be directly compared. However, it is readily seen that the dependence on the initial length, as far as tension development is concerned, differs very considerably between muscles of different design. (a) Bumblebee (*Bombus*) flight muscle (Boettiger, 1957). (b) Locust (*Schistocerca*) flight muscle (Weis-Fogh, 1956). (c) Frog sartorius muscle 0 °C (Wilkie, 1956). (d) Snail (*Helix*) pharynx retractor muscle 16 °C (Abbott & Lowy, 1956).

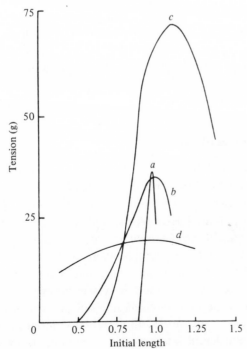

of the membrane is depolarised and more and more of the fibre is induced to contract.

Although a few vertebrate muscles have the same system as described above, most vertebrate muscle fibres are activated in an all or nothing manner, that is to say they are either fully 'switched on' or fully 'switched off' and there is no part activation. In this case, the variation in contractile force is obtained by recruiting different numbers of muscle fibres or more precisely, different numbers of motor units. At this point it is necessary to explain what is meant by a motor unit. A motor unit consists of all those muscle fibres that are innervated by the same motoneuron. All these fibres seem to have the same characteristics and they are all activated when their motoneuron conducts an action potential. The average size of motor units varies from muscle to muscle. In small muscles in which fine control is required, e.g. the extra ocular muscles, the motor units are very small – each muscle fibre has its own motoneuron. In large limb muscles which are adapted for large powerful movements only a course control is needed and in this case each motoneuron may supply as many as 200 muscle fibres.

As described below there are different kinds of motor units and therefore this system has the advantage that it is possible to recruit different kinds of muscle fibres within the same muscle, for different functions. The reason why the method of gradation of contraction is different in the vertebrates to that of the invertebrates seems to be a question of size. Invertebrates are usually much smaller than vertebrates, therefore in general they are not able to use the motor unit system as it requires reasonably large numbers of muscle fibres.

Activation of the contractile system

As mentioned, a potential difference is established across the muscle membrane. In some muscle fibres (non-twitch) there is a partial depolarisation whilst in others (twitch) it is a complete depolarisation and constitutes a propagated action potential. However in any event there has to be a link between the plasma membrane potential and the initiation of the contractile process. In many muscles the potential is taken into the fibres by an extension of the plasma membrane called the transverse tubular system. This system comes into close proximity to another system known as the sarcoplasmic reticulum. On activation the sarcoplasmic reticulum releases calcium ions which diffuse into the myofibrils. In vertebrate muscles and many invertebrate muscles the calcium interacts with a protein complex known as the troponin complex (Fig. 2). This is connected to another protein – tropomyosin and a conformational change is believed to take place on interaction with calcium ions whereby the troponin pulls the tropomyosin to one side allowing the myosin cross-bridges to interact with the actin. This form of control is referred to as I filament control as the calcium-binding proteins are on the thin or I filaments. In several species of invertebrates the switching-on mechanism seems to be in the thick (A) filaments and some of the light chains (5, 5′–dithiobis–2–nitrobenzoic acid, DTNB light chains) seem to be responsible. Many invertebrate groups seem to have both A and I filament control.

Energy consumption of different types of fibres

One of the reasons why there are different muscle fibre types is to keep energy consumption in a particular activity to a minimum. In order to achieve this, the sliding filament mechanism has been modified in various ways.

Energy consumption of different types of muscles may be measured using heat production or biochemical changes and these may be measured whilst the muscles are performing different kinds of contraction. That is to say, during isometric (force developed without any shortening), isotonic (force developed whilst shortening) and negative contractions (muscle stretched whilst developing force). All these types of contraction are involved in locomotion and it is very important that we understand something about the energy turnover under these different conditions.

Isometric contractions

It is known from work on frog muscle that isometric contractions are usually less costly than isotonic contractions (Abbott, 1951). It is also known that more energy is used during the first few seconds of an isometric contraction when force is being developed than during the latter part of a contraction when force is merely being maintained (Aubert, 1956). This means that when isometric contractions or semi-isometric contractions can be used in locomotion, for example gliding flight in birds, a considerable energy saving is achieved. The reason why isometric tension is less costly than isotonic contraction is not known. However, it is probable that more active sites become available when the thin filaments are moving past the thick filaments as they do in isotonic contractions. The initial burst of energy utilisation in an isometric contraction is most probably due to some filament movement.

When expressing the energy expenditure of a muscle during isometric contraction, it is necessary to use the term economy rather than efficiency (efficiency, strictly speaking, is the amount of available energy that is converted into work and as there is no shortening during an isometric contraction, the muscles do not perform external work). Economy is the amount of energy used for maintaining a given amount of tension for a certain length of time. In other words, it is the time–tension integral divided by the amount of energy. Experiments in comparative muscle energetics have shown that some muscles, usually slow contracting muscles, are very economical at maintaining isometric tension. Table 1 shows some measurements carried out on fast and slow muscles of the hamster. From this table it will be seen that the economy of the slow muscle is almost three times greater than that of the fast muscle. The greater economy of the slow muscle can be fairly well understood if one considers what is happening at the cross-bridge level. In the slow muscle the cross-bridge cycle is longer; this means that each cross-bridge will have to be reprimed with ATP fewer times per second than is the case in fast muscle fibres. ATP is thus used only when the cross-bridge detaches; whilst the cross-bridges are engaged and pulling they are not using ATP. Hence, the longer the engagement time of the cross-bridges the more economically they can maintain tension. Some tonic muscles such as the chicken anterior latissimus dorsi, which is the muscle which holds the wings back against the body, can maintain tension for long periods of time with little usage of energy (Goldspink, Larson & Davies, 1970; Matsumoto,

Table 1. *The expenditure and economy of developing and maintaining isometric tension by the hamster biceps brachii (fast) and soleus (slow) muscles*

	Muscle type	Length of contraction (s)		
		1.2	30	60
Expenditure (μmol high energy phosphate g^{-1})	Fast	1.5 ± 0.06	2.2 ± 0.05	2.5 ± 0.05 } *in vitro*
	Slow	1.1 ± 0.03	1.5 ± 0.02	1.8 ± 0.01
	Fast	—	—	2.2 ± 0.01 } *in vivo*
	Slow	—	—	1.1 ± 0.01
Economy gs (μmol^{-1}) high-energy phosphate)	Fast	99 ± 7	2483 ± 110	3635 ± 180 *in vitro*
	Slow	143 ± 8	6377 ± 250	9012 ± 230
	Fast	—	—	499 ± 431 } *in vivo*
	Slow	—	—	9963 ± 560

From Awan & Goldspink (1972).

Hoekman & Abbot, 1973; Goldspink, 1975). However, these muscles are extremely slow contracting tonic muscles and it is unlikely that they are ever used for any function other than a strictly postural one.

Isotonic contractions

In isotonic contractions, the muscles are shortening and hence doing work. This type of activity is seen in terrestrial animals in swinging the limbs forwards and backwards, in fish in flexing the trunk from side to side and in birds in moving the wings up and down.

As muscle is a machine capable of doing work, the question of its efficiency naturally arises. Contractile efficiency measurements are, in practice, very difficult to make because there are so many other parameters that must be standardised, particularly if one is trying to measure maximum values. Kushmerick & Davies (1969) found the efficiency of frog muscle working at optimum velocity to be over 66%. This figure refers to the efficiency of the contractile process plus the ion pumping involved in activation. These authors considered that an appropriate value for the contractile process itself might be of the order of 80–90% if the energy used for the ion pumping was subtracted from the total. It seems therefore that the transduction of energy at the cross-bridge level is probably extremely efficient given the right conditions. An even higher value of 77% for the contractile efficiency plus activation has been given for tortoise muscle by Woledge (1968). However, he found that the overall efficiency which includes the replenishment reactions, was about 35%. Tortoise muscle, therefore, appears to be considerably more efficient than frog muscle. This raises the interesting possibility that during the course of evolution it has in some cases been possible to sacrifice speed for efficiency.

The author and co-workers have studied the energetics of fast and slow muscle fibre by choosing suitable mammalian and avian muscles in which there is a preponderance of one fibre type. In order to find the maximum efficiency of these muscles, experiments have been carried out in which muscles treated with iodoacetate/cyanide were made to contract isotonically whilst loaded with different weights. The energy used was estimated by measuring the amount of phosphoryl creatine used and the mechanochemical efficiency was calculated from the work done and the amount of high-energy phosphate used. The efficiency of the fast and slow muscles is then plotted over a range of shortening velocities as shown in the curves in Fig. 6 left. It will be seen that the slow muscle is working most efficiently at shortening velocities of about one muscle length per second, whilst the fast muscle is working most efficiently at a shortening velocity of five muscle

lengths per second (the dotted line is for hypothetical muscles with intermediate rates of shortening). It will also be seen that, at its optimum rate of
shortening, the slow muscle is more efficient than the fast muscle.

The reason why slow muscles have a higher maximum efficiency is not
known. It is likely that the cross-bridges in the slow muscles are able to make
a more complete stroke and therefore they do more work per molecule of
ATP used. In addition to plotting efficiency values it is also useful to plot the
rate of energy turnover over the same range of shortening velocities (Fig. 6
below). From this plot it is seen that the rate of energy turnover is dependent
on shortening velocity. Energy utilisation is low when there is no movement
of the filaments (isometric contraction) however it increases as the rate of
shortening increases until the optimum velocity is exceeded when the energy
utilisation decreases again. It is apparent that when the thin filaments are
moving too rapidly very few of the cross-bridges are able to make contact
and complete their cycle therefore very little ATP is used. These considerations of efficiency and energy turnover have obviously been very
important in the design of muscles for locomotion.

Fig. 6. The efficiency and rate of high-energy phosphate utilisation by
mammalian fast, glycolytic (FG) and slow, oxidative (SO) muscles. The
hypothetical position of the fast oxidative glycolytic (FOG) fibres is
shown as a dotted line. The efficiency value is based on the assumption
that each mole of phosphoryl creatine split is equivalent to 468.6 J. The
rate of splitting of high-energy phosphate is in n moles s^{-1} per μmole of
total creatine.

These plots demonstrate why different types of fibres are required if
the muscle is to contract reasonably efficiently over a range of shortening
velocities.

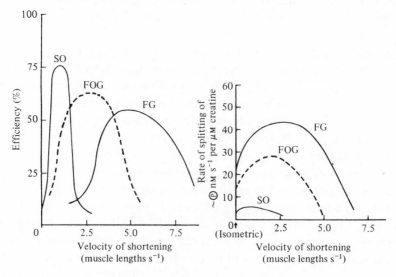

Negative work contractions

The situation often arises during locomotion that muscles are stretched whilst still developing tension (Fig. 7). When this is the case, the muscles usually develop more tension than they would do during an isometric contraction. From the energetics point of view, this extra tension is believed to be associated with no extra energy cost and this has important implications as far as locomotion is concerned. It means that we can descend a rope or a flight of stairs with less effort than it takes to ascend, even though our muscles have to develop approximately the same force whether we are going up or down. The fact that negative work requires less total energy than positive work was demonstrated in a very interesting way by Abbott, Bigland & Ritchie (1952). They arranged to have two bicycle ergometers connected back to back. One person (a large male athlete) pedalled in the normal forward direction whilst the other person (a female athlete) pedalled backwards and resisted the action to a measured extent. They showed that the female doing negative work on the double bicycle could easily resist the efforts of a large healthy male athlete doing positive work. Thus, there seems to be little doubt therefore that negative work is considerably less expensive

Fig. 7. The ability to do negative work is an important feature of the sliding filament theory. Left – shows the extension of the typical force–velocity plot (—) into the negative work region. (— — power curve). Right – shows the additional force obtained when a muscle is stretched whilst it is contracting. This extra force (shaded region) is believed to be due to pulling out the cross-bridges and therefore negative work as a whole is less costly than positive work.

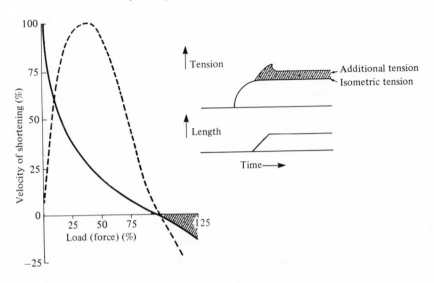

than positive work. Margaria (1968) demonstrated the importance of negative work contractions in human locomotion by measuring the energy cost of walking uphill and downhill. He concluded that in walking on the level that negative work was an important component but that it required relatively little energy.

This ability of muscle to do negative work relatively cheaply can be demonstrated on muscles working *in vitro* and indeed it is one of the inherent characteristics of the sliding filament/cross-bridge type of mechanism. Recently, Curtin & Davies (1975) have shown that when the frog sartorius muscle is stretched at about 0.18 muscle lengths per second whilst it is developing force, the rate of ATP usage related to tension is only about 25% of that during an isometric contraction (Table 2). The reason why less energy is used per given amount of tension developed seems to be that extra tension is obtained by the pulling out of the cross-bridges. It is known that there is quite a lot of compliance in the cross-bridges. The hinge part is a double helix and extensible and the S1 or head part is believed to rotate before it is pulled off the actin filament (Huxley, 1974). It seems fortunate from the point of

Table 2. *Rates of total ATP breakdown by FDNB-treated muscles during stretching and shortening (negative work)*

Velocity (muscle lengths, l_o, per second)	n	Mean rate of total chemical change (μmol g^{-1} s^{-1})
A. Isometric contraction		
0.00	7	$+0.72 \pm 0.09$
B. Stretching		
0.13	11	$+0.18 \pm 0.042$
0.18	10	$+0.18 \pm 0.042$
0.33	7	$+0.22 \pm 0.143$
0.66	18	$+0.49 \pm 0.163$
1.33	10	$+0.45 \pm 0.173$
2.00	18	$+0.84 \pm 0.351$
C. Shortening		
0.20	10	$+0.69 \pm 0.035$
0.33	13	$+0.75 \pm 0.072$
0.66	16	$+1.08 \pm 0.105$
0.91	9	$+1.41 \pm 0.099$
1.33	9	$+0.73 \pm 0.065$
2.10	10	$+0.94 \pm 0.149$

From Curtin & Davies (1975). FDNB: 2, 4, fluorodinitrobenzene.

view of locomotion, particularly terrestrial locomotion, that muscle can develop extra tension when stretched in the activated state and that the tension produced during negative work is very inexpensive.

Contractile properties of different kinds of vertebrate muscle fibres

Only striated muscle is used for vertebrate locomotion; this still shows considerable diversity. There are several classifications used for muscle fibres however. The scheme outlined below is, I believe, the most informative.

Tonic muscle fibres

These fibres are very slow contracting fibres which do not usually exhibit a propagated muscle action potential, hence when stimulated with a single stimulus they do not produce a significant response. They are multiply innervated and have a graded response to stimulation of different frequencies. Tonic fibres have a very long cross-bridge cycle and therefore a low intrinsic rate of contraction. These fibres are capable of maintaining isometric tension very economically and are therefore used for maintaining posture. As the tonic fibres are very slow in contracting, they are not normally involved in locomotory movements.

Slow phasic fibres (SO fibres)

These fibres are slow contracting but they are not as slow as the tonic muscle fibres. They have a propagated action potential, hence they are referred to as twitch fibres. This type of muscle fibre is often innervated with a single end plate although in some animals, e.g. fish, they may be multiply innervated. Slow phasic fibres have a reasonably slow cross-bridge cycle, they use ATP at a slow rate and they are both economic in maintaining isometric tension and efficient in doing work providing they are contracting slowly. They usually contain a lot of mitochondria and hence they are resistant to fatigue. They are therefore used for maintaining posture (isometric tension) and for carrying out slow repetitive movements.

Fast, phasic, glycolytic fibres (FG fibres)

These fibres are adapted for a high power output and they have a rather poor thermodynamic efficiency for producing work. They have a very short cross-bridge cycle time and hence a high intrinsic speed of shortening.

Fast phasic fibres are usually recruited only when very rapid movement is required. They usually possess very few mitochondria, because a very large proportion of the fibre volume would have to be mitochondrial in order to supply ATP as fast as it is used and this would be incompatible with a high tension and hence a high power output. They therefore fatigue very rapidly and most of the replenishment of their energy supplies takes place after the exercise has ceased. During contraction the energy that is supplied comes mainly from glycolysis.

Fast, phasic, oxidative fibres (FOG fibres)

This kind of fibre is essentially the same as the fast glycolytic fibre except that they contain reasonably large numbers of mitochondria. In some cases, but not in all, they may have a slightly lower intrinsic rate of contraction than the fast glycolytic fibres. These fast oxidative fibres are apparently adapted for reasonably fast movements of a repetitive nature.

The different kinds of muscle fibres can be distinguished histochemically (Fig. 8) by staining for myofibrillar (myosin) ATPase. This is most effective

Fig. 8. Three types of fibres of vertebrate striated muscle using the histochemical myosin ATPase method (Ward, 1976).

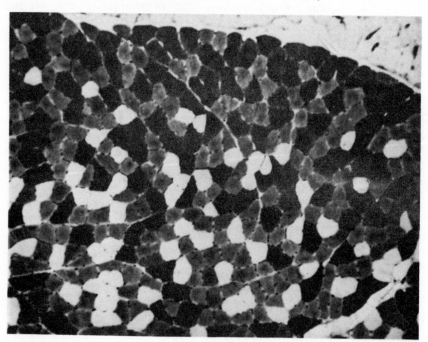

if it is combined with pretreatment of the section at acid or alkaline pH. Myofibrils (myosin) of the fast type are known to be reasonably alkali stable and therefore after treatment at say pH 10.4, the fast fibres still stain darkly. The myofibrils of the slow fibres are more acid resistant therefore when preincubation is carried at pH 4.3 it is the slow fibres which stain darkly. Oxidative enzyme stains are also useful, e.g. succinic dehydrogenase, particularly when these are combined with the ATPase stain using alternate sections. In this way it is possible to distinguish the fast glycolytic fibres from the fast oxidative fibres with reasonable certainty.

Muscle fibre composition and different kinds of locomotion

Fish Swimming

The division of labour between the different types of fibres is perhaps best exemplified by the role of the different types of muscle in fish swimming. Fig. 9 shows the generalised diagram of distribution of fibre types in fish. Studies using electromyography (Bone, 1966; Hudson, 1972; Davison, Goldspink & Johnston, 1976) and biochemical studies (Bone, 1966; Johnston & Goldspink, 1973) have indicated that the red muscle fibres, usually found near to the lateral line, are the ones which are

Fig. 9. Diagram of fish muscle showing the positions and recruitment of the red (SO) pink (FOG) and white (FG) fibres. EMG electrodes were inserted at the three layers and recordings were made at different swimming speeds. From Johnson *et al.* (1977).

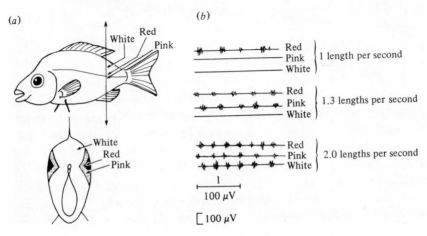

responsible for slow cruising movements. Measurements of the specific activity of the myosin ATPase of the different fibre types have shown these fibres to be slow contracting fibres (Johnston, Frearson & Goldspink, 1972). Energetically, this makes sense because the slow (SO) fibres will work with high efficiency at the sort of speeds involved in slow cruising. The slow muscle fibres may only account for a few per cent of the total musculature. However, the power requirement for slow swimming is very low. The white muscle fibres have a fast myosin ATPase and these are believed to be recruited only when rapid bursts of activity are required. The white (FG) fibres account for 70–80% of the musculature and it seems paradoxical that all this muscle is inactive most of the time. However, unlike the situation in the bird, this additional weight is no great problem to the fish because of its buoyancy in the aqueous environment. The precise function of the pink (FOG) muscle fibres is not known; however, from biochemical studies (Johnston, Davison & Goldspink, 1977) it seems that this muscle has an intrinsic speed of shortening that is intermediate between that of the red and the white fibres. Thus, it may be assumed that the pink muscle fibres are recruited for swimming speeds which are slightly faster than slow cruising. These three kinds of fibres with different optimum rates of shortening enable the fish to swim at a range of speeds without a marked drop in efficiency. The fish, therefore, has in a sense, a three-geared system.

The involvement of the different kinds of muscle fibre does not require a complicated recruitment system by the central nervous system. The red fibres will no doubt be recruited first, the pink fibres will be next and if the speed of swimming has to be increased further then the white muscle fibres will become involved. At high speed the pink and particularly the red fibres will be working inefficiently. However, there is no need for these to be turned off by the central nervous system because they become mechanically ineffective at the higher speeds and they in fact split very little ATP (see Fig. 6 page 12).

The red muscle fibres are well supplied with mitochondrial enzymes and obviously have a predominantly oxidative metabolism. They probably always work in 'steady state' and never experience an oxygen debt. The pink fibres also have fairly high levels of oxidative enzymes but they are also particularly well supplied with glycolytic enzymes. They are, however, reasonably fast fibres, with a fairly high energy turnover and they apparently do experience some oxygen debt. However, because the pink fibres are recruited reasonably frequently they also require the ability to recover rapidly, thus it is seen that they are generally very metabolically active. White muscle fibres, on the other hand, will consume energy at a very rapid rate when activated and there is no way that the fish could get sufficient oxygen to this tissue to maintain it in a steady state. There is little point, therefore, in these fibres having high levels of oxidative enzymes. Because these white fibres fatigue very quickly, they are only used for short periods of time and most of the replenishment of immediate energy supplies and the oxidation of the lactate has to take place after the cessation of activity. Fish therefore afford a good example of the way that the energetic requirements for locomotion can be met using different types of muscle fibres.

Bird flight

The situation in birds is rather different as the muscle fibre composition of the flight muscles is fairly homogeneous. It seems that the power requirements of flight are such that the bird cannot apparently indulge in a two- or three-geared system. The pectoral muscles in many birds make up 40% of the body weight and any diversity in the fibre population would add to this weight. In general we find that each species of bird has a particular flapping frequency and that this frequency does not vary to any great extent during flight. The intrinsic speed of shortening of the fibres is therefore matched to the basic flapping frequency of the bird. The flapping frequency of different species is related to body size (Greenwalt, 1962). In general, this means that the larger the bird the slower its muscle fibres; although they are still phenotypically fast fibres, i.e. they have three light chains (Talesara & Goldspink, 1978).

The situation is rather different for gliding flight as the muscle action is basically isometric. Therefore, slow muscles will be a distinct advantage as they can produce isometric tension more economically than fast muscle fibres. This is probably one of the reasons why large birds have been able to use gliding as a means of energy saving. The energy saving during gliding has been shown to be very considerable (Baudinette & Schmidt-Nielsen, 1974). The reason for this is that the force required during gliding is only that required to maintain the posture of the bird in the air and so only a relatively few muscle fibres (motor units) need to be recruited (Goldspink, Mills & Schmidt-Nielsen, 1978). The other important factor in energy saving is that muscle activity in gliding flight is predominantly isometric and the energy turnover by the contractile system will be low in the few fibres that are activated. Although the pectoral muscles are fairly homogeneous there is some evidence that in the herring gull there is a small percentage (8%) of slow phasic fibres and no doubt these are used for gliding (Talesara & Goldspink, 1978).

Walking and running in terrestrial vertebrates

Our ideas about muscle action during walking and running have changed over the past ten years or so. The work of McNeill Alexander (1977), Taylor *et al*. (1974), Cavagna (1977) and McMahon (1977) has shown that in running a considerable amount of energy is stored from step to step as elastic strain energy in the tendons. This involves an appreciable change in the length of the tendons rather than in the muscle itself. The function of the muscles seems to be to tension the tendons and this means that the group of muscles involved in this activity will be developing mainly isometric force (tension). As isometric contractions are considerably less costly than isotonic contractions, this presumably results in quite an energy saving. The animals are essentially bouncing along on their tendons and in a sense it is one of the ways that terrestrial animals have been able to utilise the effects of gravity. Isotonic contraction is also involved in swinging the limbs forwards and backwards. This action does not however seem to require much energy because the energetic cost of running in animals with light limbs is about the same as for ones with heavier limbs (Taylor *et al*., 1974). The muscles responsible for swinging the limbs seem to be long and parallel fibred with a large number of sarcomeres in series thus giving a high overall rate of shortening. In contrast the muscles involved in tensioning the tendons are pennate muscles which give a large effective cross-sectional area and this makes them ideal for developing large isometric forces. It is known that the cost of locomotion for large animals is considerably less than for small animals (Schmidt-Nielsen, 1971; Taylor *et al*., 1974). The reason why this is so is probably a combination of factors. One is that the intrinsic rate of contraction of the fibres is matched to the stride duration and this means that the muscle fibres in the larger animals will be slower contracting than those in the smaller animals. Slower contracting fibres, it has been shown, are both more economical and more efficient. The longer stride length means that when tensioning the tendons the longer animal will have to maintain the tension for a longer time, however they have to develop tension less frequently and it is development of tension that is the expensive part of the isometric contractions.

We may assume that in all types of terrestrial locomotion there is the same sort of progressive recruitment of fibres that is found in fish swimming. Certainly most mammalian muscles are mixed and contain populations of the three main types of fibre. Evidence for the progressive recruitment of fibre type has been obtained for human muscle by Gollnick, Piehl & Saltin (1974) and Gollnick, Karlsson, Piehl & Saltin (1974). These workers studied glycogen depletion in different types of fibres in human muscle during isometric and isotonic exercise. They found that the slow fibres were recruited first in both kinds of exercise. For example in bicycling at slow speeds only the slow fibres were used but as the rate of pedalling was increased the fast fibres also began to be recruited. The different efficiency maxima and the progressive recruitment of the fibre types enables them to work efficiently over a wide range of shortening velocities.

Temperature adaptation in the sliding filament mechanism

Muscles in endothermic animals work over a restricted temperature range but those in some ectothermic animals have to work at more extreme temperatures. Antarctic fish, for example, experience environmental temperatures of $-1\,°C$ to $4\,°C$ whilst at the other end of the scale, fish in the hot springs of the Rift Valley in East Africa live at temperatures of $35\,°C$ to $38\,°C$. How is it that the Antarctic fish can move about at these low temperatures? Recent work by the author and co-workers has shown the contractile apparatus of Antarctic fish differs from that of warm-water or even-temperature-water fish in that it is designed to have a high specific ATPase activity at the low temperature range (Johnston, Walesby, Davison & Goldspink, 1975). The increased specific activity seems to be associated with a difference in the tertiary structure of the proteins because the myofibrils are very susceptible to heat denaturation (Fig. 10). At higher environmental temperatures fish require a fairly heat-stable contractile system, therefore these fish have evolved myosin with a more rigid type of molecular structure.

Many species of fish live within a very restricted temperature range, and hence they have evolved a myofibril system which is adapted for this range. However, a few species of fish are able to adapt to a wide range of temperatures; the carp family provides several examples of fish with this ability. For instance the common goldfish can survive and move around reasonably well at temperatures of about $+1\,°C$ in an ice-bound garden pool, whilst on the other hand it can live quite happily in an ornamental pool in the tropics. The carp family can acclimate to these temperatures because it has the ability to alter the characteristics of the contractile proteins. During acclimation to a low environmental temperature they produce a myosin with an ATPase which is active at the lower end of the temperature scale but which is readily denaturated by heat. Fish acclimated to a high environmental temperature have a myosin ATPase that is only active at the higher temperatures but which is heat stable (Johnston et al., 1975) (Fig. 10).

Alterations in the rate of muscular contraction during temperature acclimation may involve changes in the rate at which the cross-bridges work (myosin ATPase activity) but they may also involve the rate at which the active sites become available on the actin filaments. Work by Dr Ian Johnston at the University of St Andrews (Johnston, 1979) and in our own laboratory has shown that the troponin complex changes during temperature acclimation. Indeed if the troponin is removed from myofibrils prepared from warm- and cold-adapted fish the myofibrils no longer show the elevated ATPase activity at low temperatures. If the troponins are then replaced, a difference is again obtained. It seems therefore that the troponin complex in fish muscle may be involved in determining the rate of cross-bridge activity rather than acting as just a simple on/off switch.

In this paper I have attempted to present reasons why the sliding filament mechanism has been so widely adopted throughout the animal kingdom. As well as being an effective and efficient mechanism for producing and main-

Fig. 10. Plots of ATPase activity (mol Pi mg^{-1} min^{-1}) (left) and rate of temperature denaturation (right) for myofibrils from Antarctic and tropical fish (top) and warm and cold acclimated fish (bottom). Data from Johnston *et al.* (1975), and Johnston, Davison & Goldspink (1975).

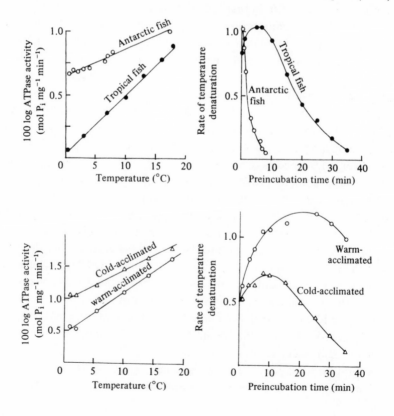

taining force it is also a mechanism which has lent itself to a considerable amount of modification.

References

Abbott, B. C. (1951). The heat production associated with the maintenance of a prolonged contraction and the extra heat produced during large shortening. *J. Physiol., Lond.* **112**, 438–45.

Abbott, B. C., Bigland, B. & Ritchie, J. M. (1952). The physiological cost of negative work. *J. Physiol., Lond.* **117**, 380–90.

Abbott, B. C. & Lowy, J. (1956). Resting tension in snail muscle. *Nature, Lond.* **178**, 147–8.

Alexander, R. McN. (1974). The mechanics of jumping by a dog (*Canis familiaris*). *J. Zool., Lond.* **173**, 549–73.

Alexander, R. McN. (1977). Terrestrial locomotion. In *Mechanics and Energetics of Animal Locomotion*, ed. R. McN. Alexander & G. Goldspink, p. 168. London: Chapman & Hall.

Aubert, X. (1956). *Le Couplage Energetique de la Contraction Musculaire*. Brussels: Editions Arscia.

Awan, M. Z. & Goldspink, G. (1972). Energetics of the development and maintenance of isometric tension by mammalian fast and slow muscle. *J. Mechanochemistry, cell motility*, **1**, 97–108.

Baudinette, R. V. & Schmidt-Nielsen, K. (1974). Energy cost of gliding flight in herring gulls. *Nature, Lond.* **248**, 83–4.

Boettiger, (1957). In *Recent Advances in Invertebrate Physiology*, ed. B. T. Sheer, p. 117. University of Oregon Publishers.

Bone, Q. (1966). On the function of the two types of myotomal muscle fibres in elasmobranch fish. *J. mar. biol. Ass. UK*, **46**, 321–49.

Calow, L. J. & Alexander, R. McN. (1973). A mechanical analysis of a hind leg of a frog. *J. Zool., Lond.* **171**, 293–321.

Cavagna, G. A. (1977). Aspects of efficiency and inefficiency of terrestrial locomotion. In *Biomechanics*, vol. VI A, ed. E. Asmussen & K. Jørgensen, p. 3. Baltimore: University Park Press.

Curtin, N. A. & Davies, R. E. (1975). Very high tension with little ATP breakdown by active skeletal muscle. *J. Mechanochemistry, cell motility*, **3**, 147–54.

Davison, W., Goldspink, G. & Johnston, I. A. (1976). Division of labour between fish myotomal muscles during swimming. *J. Physiol., Lond.* **263**, 185P.

Goldspink, G., Larson, R. E. & Davies, R. E. (1970). Thermodynamic efficiency and physiological characteristics of the chick anterior latissimus dorsi muscle. *Z. vergl. Physiol.* **227**, 848–50.

Goldspink, G. (1975). Biochemical energetics for fast and slow muscles. In *Comparative Physiology: Functional Aspects of Structural Materials*, ed. L. Bolis, H. P. Maddrell & K. Schmidt-Nielsen, pp. 173–85. Amsterdam: North-Holland.

Goldspink, G., Mills, C. & Schmidt-Nielsen, K. (1978). Electrical activity of the pectoral muscles during gliding and flapping flight in the herring gull. (*Larus* argentatus). *Experientia*, **34**, 862.

Gollnick, P. D., Karlsson, J., Piehl, K. & Saltin, B. (1974). Selective glycogen depletion in skeletal muscle fibres of man following sustained contraction. *J. Physiol., Lond.*, **241**, 59–67.

Gollnick, P. D., Piehl, K. & Saltin, B. (1974). Selective glycogen depletion patterns in human skeletal muscle fibres after exercise of varying intensity and at varying pedalling rates. *J. Physiol., Lond.* **241**, 45–57.

Greenwalt, C. H. (1962). Dimensional relationships for flying animals. *Smithson. misc. Coll.* **144**, 2.

Hoyle, G. & Smyth, T. (1963). Giant muscle fibres in a barnacle *Balanus nubilis* Darwin. *Science, NY,* **139**, 49–50.

Hoyle, G., McAlear, J. H. & Selverston, J. H. (1965). Mechanism of supercontraction in a striated muscle. *J. cell. Biol.* **26**, 621–40.

Hudson, R. C. L. (1972). On the function of the white muscles in teleosts at intermediate swimming speeds. *J. exp. Biol.* **58**, 509–22.

Huxley, A. F. (1974). Muscular contraction. *J. Physiol., Lond.* **243**, 1–43.

Huxley, A. F. & Niedergerke, R. (1954). Structural changes in muscle during contraction. Interference microscopy of living muscle fibres. *Nature, Lond.* **173**, 971.

Huxley, H. E. & Hanson, J. (1954). Changes in the cross-striations of muscle during contraction and stretch and their structural interpretation. *Nature, Lond.* **173**, 973.

Jahromi, S. S. & Atwood, H. L. (1969). Correlation of structure speed of contraction and total tension in fast and slow abdominal muscle fibres of the lobster (*Homarus americanus*). *J. exp. Zool.* **171**, 25–38.

Johnston, I. A. (1979). Calcium regulatory proteins and temperature acclimation of actomyosin ATPase from a Eurythermal Teleost (*Carassius auratus* L.). *J. comp. Physiol.* **129**, 163–7.

Johnston, I. A., Davison, W. & Goldspink, G. (1975). Adaptations in Mg^{2+}... activated myofibrillar ATPase activity induced by temperature acclimations. *FEBS letters,* **50**, 293–5.

Johnston, I. A., Davison, W. & Goldspink, G. (1977). Energy metabolism of carp swimming muscles. *J. comp. Physiol.* **B114**, 203–16.

Johnston, I. A., Frearson, N. & Goldspink, G. (1972). The effects of environmental temperature on the properties of myofibrillar adenosine triphosphatases from various species of fish. *Biochem. J.* **133**, 735–8.

Johnston, I. A. & Goldspink, G. (1973). A study of the swimming performance of the Crucian carp *Carassius carassius* L in relation to the effects of exercise and recovery on biochemical changes in the myotomal muscles and liver. *J. Fish Biol.* **5**, 249–60.

Johnston, I. A., Walesby, N. J., Davison, W. & Goldspink, G. (1975). Temperature adaptations in the myosin of an Antarctic fish. *Nature, Lond.* **254**, 74–5.

Kushmerick, M. J. & Davies, R. E. (1969). The chemical energetics of muscle contraction II. The chemistry, efficiency and power of maximally working sartorius muscles. *Proc. R. Soc., Ser. B.*, **174**, 315–54.

Margaria, R. (1968). Positive and negative work performances and their efficiencies in human locomotion. *Int. Z. angew. Physiol. einschl. Arbeitsphysiol.* **25**, 339–51.

Matsumoto, Y., Hoekman, T. & Abbott, B. C. (1973). Heat measurements associated with isometric contraction in fast and slow muscles of the chicken. *Comp. Biochem. Physiol.* **46A**, 785–97.

McMahon, T. A. (1977). Scaling quadrupedal galloping frequencies, stresses and joint angles. In *Scale Effects in Animal Locomotion*, ed. T. J. Pedley, p. 143. London & New York: Academic Press.

Peachy, L. D. (1965). The sarcoplasmic reticulum and transverse tubules of the frog's sartorius. *J. Cell Biol.* **25**, 209–31.

Pringle, J. W. S. (1972). Arthropod muscle. In *The Structure and Function of Muscle*, 2nd edn, vol. 1, ed. G. H. Bourne, p. 149. London & New York: Academic Press.

Pringle, J. W. S. (1975a). *Insect Flight: Oxford Biology Reader 52.* Oxford University Press.

Pringle, J. W. S. (1975b). Insect fibrillar muscle and the problem of contractility. In *Comparative Physiology, Functional Aspects of Structural Materials*, ed. L. Bolis, H. P. Maddrell and K. Schmidt-Nielsen, p. 139. Amsterdam: North-Holland.

Schmidt-Nielsen, K. (1971). Locomotion: energy cost of swimming, flying and running. *Science, NY,* **177**, 222–6.

Talesara, G. L. & Goldspink, G. (1978). A combined histochemical and biochemical study of myofibrillar ATPase in pectoral, leg and cardiac muscle of several species of bird. *Histochem. J.* **10**, 695–710.

Taylor, C. R., Shkolnik, A., Dmiel, R., Baharar, D. & Borat, A. (1974). Running in cheetahs, gazelles and goats: energy cost and limb configuration. *Am. J. Physiol.* **227**, 848–50.

Ward, P. S. (1976). Ph.D. Thesis, University of Hull.

Warner, G. F. & Jones, R. (1976). Leverage and muscle type in crab chelae (Crustacea: Brachyura). *J. Zool., Lond.* **180**, 57–68.

Weis-Fogh, T. (1956). Tetanic force and shortening in locust flight muscle. *J. exp. Biol.* **33**, 668–84.

Wilkie, D. R. (1956). The mechanical properties of muscles. *Br. med. Bull.* **12**, 177–82.

Woledge, R. C. (1968). The energetics of tortoise muscle. *J. Physiol., Lond.* **197**, 685–707.

J. D. CURREY

Skeletal factors in locomotion

The musculoskeletal system of an animal may be thought of as those parts of its body whose function is primarily mechanical. In this chapter I shall not discuss the active elements: muscles and other contractile proteins, nor hydrostatic skeletons, nor those skeletal elements, such as the skull, that are not used in locomotion.

Skeletal elements are of four main types:

(*a*) Rigid elements for interacting with the environment.

(*b*) Flexible links between these elements.

(*c*) 'Tendons' to move these rigid elements, being usually moved themselves by muscles.

(*d*) Energy-storage devices.

Rigid elements

Animals moving around frequently make use of rigid skeletal elements which interact with the ground or the surrounding fluid. Often this interaction is best carried out by fairly rigid elements. This is not always true of structures that have hydro- or aerodynamic functions. For instance, the primary feathers of a bird are able to propel the bird without any stalling only because they can twist about their long axis. However, bones of the axial skeleton and the limb bones of vertebrates, the limb elements of walking arthropods, the spines of sea urchins and the valves of swimming and burrowing bivalve molluscs all function well because they are stiff rather than pliant.

How can an element have the required stiffness? Consider an idealised but not unrealistic case. A limb element is required to support a load W over a length L and deflect by an amount z (Fig. 1). Suppose for simplicity's sake that the element has a uniform cross-section (see McMahon & Kronauer, 1976, for an interesting discussion of trees, in which uniformity of cross-section certainly does not hold). It can easily be shown that, whatever the values of L and W, deflection will be inversely proportional to IE, where I is the second moment of area and E is Young's modulus of elasticity. How can I be maximised? For a cross-section of arbitrary shape, bending about the centroid of the section, I is given by $I = \Sigma y^2 \delta A$ where y is the distance from the centroid. To maximise the value of I for a given cross-sectional area A, then crudely, the value of Σy^2 must be maximised. So the material of the element should be as far as possible from the centroidal axis, the so-called neutral axis. Ideally, as shown in Fig. 2a the element should be made of two very thin sheets a long way from the neutral axis. However, the element made of two thin sheets would be stiff only if the sheets were kept far apart. So in practice a web would have to be put between the sheets (Fig. 2b). This

Fig. 1. The limb element is designed to have the minimum mass that will deflect only by an amount z while carrying a load W over a length L.

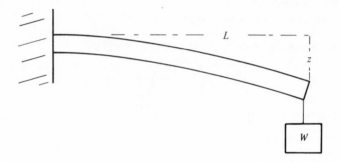

Fig. 2. (a) Two thin sheets maximise I. (b) A web takes the shear stresses and keeps the sheets apart. (c) The best arrangement if the beam can be bent in either of two directions at right angles. (d) The hollow cylinder, which bears bending in any direction, and resists torsion well. (e) The thick-walled cylinder, which will not collapse by local buckling.

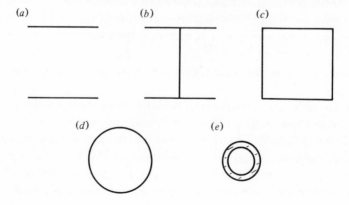

arrangement assumes that the limb element is always loaded in exactly the same plane. In reality limb elements have to be stiff in many directions. If loaded in two directions at right angles the best arrangement would be as in Fig. 2c, with two sides always acting as webs for the other two sides that act as flanges. If the element is to be loaded in any direction the best arrangement is a hollow cylinder (Fig. 2d). A hollow cylinder is also the most efficient for resisting torsion. Unfortunately, such a cylinder would have such airy and unsubstantial walls that they would buckle if loaded in bending or axially. Although the value of I may be high, the thickness of the wall is too small in relation to the diameter. The ratio of wall thickness to diameter that will cause the cylinder to fail by buckling is known empirically only. A rough approximation is $D/t = E/(2 \times \text{stress})$, where E is the modulus of elasticity, and D/t is the critical value of ratio of diameter to wall thickness, and stress is the stress in the material.

Two groups of animals are overwhelmingly the most important users of hollow limb elements: the arthropods and the vertebrates. Table 1 shows the value of D/t that is likely to cause the limbs to fail by local buckling rather than rupture through reaching their maximum stress.

So, in both groups, buckling is likely to be important when the diameter approaches forty times the wall thickness. In fact, because of local imperfections in the walls, which cause premature buckling, the actual value is likely to be a good deal less than this. Although the shafts of long bones rarely approach this value, they may become very thin-walled towards their ends, but here the wall is supported by cancellous bone. Arthropod podomeres, on the other hand, are quite often very thin-walled, and this is particularly true of spiders (Currey, 1967).

Suppose we have a cylindrical long bone, adapted therefore to being bent in any direction. The deflection that a given load will impose on it is inversely proportional to its second moment of area (I). Natural selection probably produces near-minimum-mass solutions to problems of mechanical design. That is to say, in teleological terms, the limb elements are designed to have a certain stiffness or strength, and to have as little mass as possible while being

Table 1. *Modulus, maximum stress and critical value of D/t for bone and stiff locust cuticle*

	Modulus of elasticity (E/N m^{-2})	Strength (Max. stress/N m^{-2})	Critical D/t
Bone	1.4×10^{10}	1.8×10^8	39
Stiff cuticle	10^{10}	10^8	50

that stiff or strong, rather than having a certain mass and being as stiff or strong as possible with that mass. This is particularly so of animals engaged in active locomotion. For them minimum mass is very important because the greater the mass of a limb, the greater its moment of inertia about its proximal joint. The power required to accelerate and decelerate limbs can be a high proportion of the total power output of locomotion (Cavagna, Saibene & Margaria, 1964). The *weight* of the limb is also important and undesirable, particularly in those animals, such as the kangaroo, that undergo large changes in potential energy in each stride and also, of course, in swimming and flying animals.

Considering stiffness first, we can calculate the effect on total mass of altering the ratio of the internal (d) and external (D) diameters while keeping I constant (Fig. 3). The lower solid line shows that as the ratio of d/D approaches unity, the mass declines, until at some point, probably near the critical value, the danger of local buckling would make further thinning of the cortex and expansion of the whole bone unadaptive.

This would seem to show that the minimum mass for bony elements should be constrained by the local buckling criterion. However, real bones, except some bird bones, are not empty, but filled with marrow. Much of the marrow seems to have little metabolic function, serving merely to fill the cavity. This has interesting consequences, as Alexander (1975) points out.

Fig. 3. Diagram showing relationship between mass and d/D for elements with a constant value of I.

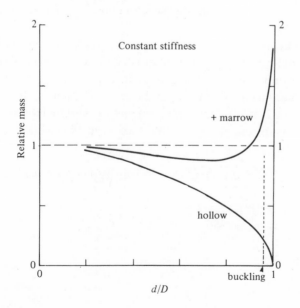

As the marrow cavity gets greater, so does the mass of marrow. The effect of taking into account the mass of marrow is shown by the upper solid line in Fig. 3. This assumes that the density of the marrow is half that of the bone, and that the marrow does not contribute to I.

(a) There is now a minimum mass at an intermediate value of d/D, about 0.7, and the mass of the bone plus marrow at this value is only 12% less than the mass of a completely solid bone of the same value of I.

(b) The curve near the minimum is very flat.

These results imply that, for hollow cylinders of bone, filled with fat, the minimum mass solution will impose quite thick walls, much thicker than are likely to fail through local buckling. Furthermore, the range of wall thicknesses producing mass values close to the minimum is quite large. Selection will not, therefore, act strongly to produce great uniformity between different bones and between different animals.

If we assume that the critical value is the minimum mass to produce a particular bending *strength* of an element, then instead of assuming a constant value for I, we assume a constant value for D/I. (The maximum stress in a beam is proportional to D/I.) Fig. 4 shows that again, if the bone is assumed empty, the declining minimum-mass curve cuts the local buckling line, but if marrow is assumed there is a very shallow minimum at $d/D\ ca\ 0.6$, with the saving in mass compared with a solid bone of equal strength of only 10%.

Fig. 4. Diagram showing relationship between mass and d/D for elements with a constant value of D/I.

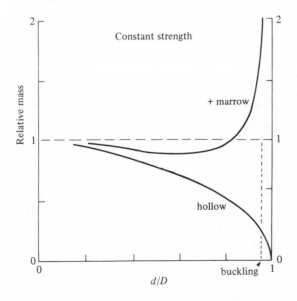

Many long bones of mammals do seem to have values of d/D in the region of 0.5–0.75.

Birds' long bones are sometimes marrow-filled, sometimes hollow; it is not always homologous bones that are hollow (Headley, 1895). It would be most interesting to examine the ratio of d/D in various bird bones, to see to what extent (if any) those without marrow were thinner-walled. Of course, in birds aerodynamic constraints require that D be not too large, so it is unlikely that any really bloated, thin-walled bones occur.

The situation in arthropods is different because at least some, and often most, of the cavity of a podomere is occupied by muscle. This muscle has to be placed somewhere, and therefore that proportion of the cavity of a podomere that is occupied by muscle is not contributing almost useless mass to the limb, as is the yellow marrow of a bone. Not surprisingly, arthropod limb elements seem to have in general higher values of d/D than long bones, taking them considerably nearer to, and sometimes beyond, the critical value of d/D for local buckling. (See Currey, 1967, for values of d/D.) Since the density of arthropod cuticle is not much greater than that of fat, hollow podomeres filled with fat would have a miniscule mass advantage over solid ones.

So far we have discussed the *build* of the skeletal elements, but have hardly considered the mechanical properties of the material itself. The fact that deformation of a beam is inversely proportional to IE implies that altering I or E are both possible ways of affecting stiffness.

The Young's modulus of bone is variable. In the red deer antler it is about $7 \, \mathrm{GN \, m^{-2}}$, in the cow's femur about $14 \, \mathrm{GN \, m^{-2}}$, and in the tympanic bulla of the fin whale it is $31 \, \mathrm{GN \, m^{-2}}$ (Currey, 1979). These differences relate to differences in the mineral content, from 59% in antler to 86% in the bulla. However, this stiffness must be paid for. In these three bone types the resistance to fracture under the influence of impact declines sharply with increasing stiffness and mineralisation. In a linearly elastic solid the energy absorbed per unit volume up to fracture is proportional to strength^2/Young's modulus. So unless an increase in modulus by a factor x is accompanied by an increase in strength of at least $\sqrt{x}$, the energy absorbed, which is at least partially a measure of impact resistance, will decline. Furthermore, the increasing mineralisation, which produces the increased modulus, can be shown to increase the ease a crack has in travelling through the material. This effect too, will make the bone more susceptible to impacts. Highly mineralised bone is, therefore, less tough than less mineralised bone. The amount of mineralisation, with its concomitant effects, that is most adaptive for a limb will depend upon the conditions in which the bone has to function.

Arthropod cuticle is much more variable in its modulus than bone. It can vary from roughly $100 \, \text{kN m}^{-2}$ in the extraordinarily pliant intersegmental membrane of female locusts (Vincent, 1975) to the extremely stiff cuticle near joint articulations, whose modulus probably exceeds $20 \, \text{GN m}^{-2}$ (Bennet-Clark, 1975). These differences in modulus are produced, not by mineralisation, but mainly by differences in the amount of cross-linking of the protein component of the cuticle. The costs associated with this high modulus are graphically illustrated by Bennet-Clark's account of how lightly scoring the outer surface of the highly stressed semilunar process of the femur of a locust results in brittle fracture when the animal tries to jump. (The experiment is easy to perform, and the fracture occurs with an unpleasant click.) The considerable degree of cross-linking makes the cuticle brittle so it cannot accommodate the locally high stresses at the base of the score line.

Calculations can be made on materials to show the best mix of properties for particular functions (Shanley, 1960; Wainwright, Biggs, Currey & Gosline, 1976). For instance, consider a short column of unit length that must support a weight P and have a minimum mass. Suppose the material of the column has a compressive strength σ_1 and the density ρ_1. A mutation arises that alters the strength to σ_2, but at the same time alters the density to ρ_2. Would such a mutation be advantageous? The cross-sectional areas will be P/σ_1 and P/σ_2, and the masses will be $(P/\sigma_1) \, \rho_1$ and $(P/\sigma_2) \, \rho_2$. As P is given, this is equivalent to saying that the mass necessary to do the job will be proportional to ρ/σ. Therefore, to decrease mass, decreasing density or increasing strength will have the same effect. This is perhaps obvious, what are less obvious are minimum-mass criteria for other loading systems of failure modes.

Table 2 concerns four modes of failure and four structures. The failure

Table 2. *The functions that must be minimised to minimise the mass of various kinds of structure, potentially failing in various ways (ρ = density; E = modulus of elasticity; σ = yield stress); the calculation for the beam assumes that its section is symmetrical about two axes at right angles e.g. is square or circular*

	Rupture	Energy absorption	Deformation	Euler buckling
Short column	ρ/σ	$E \cdot \rho/\sigma^2$	ρ/E	—
Slender column	ρ/σ	$E \cdot \rho/\sigma^2$	$\rho/\sqrt{E}$	$\rho/\sqrt{E}$
Beam	$\rho/\sigma^{\frac{2}{3}}$	$E \cdot \rho/\sigma^2$	$\rho/\sqrt{E}$	—
Tensile structure	ρ/σ	$E \cdot \rho/\sigma^2$	ρ/E	—

modes are: (a) rupture in static loading (either in tension, compression or bending); this depends on the maximum stress in the structure; (b) exceeding the energy-absorbing capacity of the material, which is important in dynamic loading; (c) deforming excessively; (d) undergoing Euler buckling (the buckling characteristic of a slender column loaded along its length).

These calculations assume that the materials are completely brittle, which is not strictly true (or even nearly true, in some cases) but does indicate when irreversible change has taken place.

The table examines four types of structure common in locomotion. The values ρ, E and σ refer to density, Young's modulus and strength respectively. The entries in the body of the table show the quantity that must be minimised to give the structure of least weight to do a particular job.

The significance of Table 2 is that it shows that, for different structures, or for the same structure loaded in different ways, different properties become more or less important, advantageous or deleterious.

For instance, in a tensile structure that must not deform too much, proportional changes in density and modulus are equivalent, whereas in a beam that must not bend too much a change in density is more important than a proportional change in modulus. Natural selection balances the competing mechanical and other costs and benefits of different material properties and arrives, as can be seen from Table 3, at some very different end products.

Flexible links

Stiff skeletal elements must move relative to each other for locomotion to occur. The joints in limbs, and between the limbs and the more rigid body, characteristically appear to have one or three degrees of freedom. (That is, to be a hinge or have the mobility possible from a ball-and-socket type joint.) The apparent simplicity often belies the real situation; for instance the human knee appears to have one degree of freedom, of flexion and extension. In fact, it has also a small amount of ability to rotate the long axes of the femur and the tibia with respect to each other, and so the knee joint has two degrees of freedom.

The two major variables that affect the design of joints are the general mobility of the joint and the force to be transmitted across it. If the joint has limited mobility it can be simple, even if the forces across it are large. For example, the hinge joints of bivalves consist of a pad of rubbery tissue between two grooves, or between a spoon shaped projection in one valve

Table 3. *The mechanical properties of a number of biological materials, mostly used in locomotion. The values are very approximate, particularly for some low-modulus materials, but they do give some idea of order of magnitude. The original references should be consulted before any strong conclusions are drawn! E = modulus of elasticity; σ_y = yield stress in tension; σ_t = ultimate stress in tension; σ_b = bending strength; ϵ = yield strain (figures with asterisks: ultimate strain); ρ = density. Many values for density are unreported, but they are likely to be near 1100 kg m⁻³. The values for the functions to be minimised have been multiplied by various factors to produce sensible numbers. The numbers in each column have been multiplied by the same factor*

	E (GN m⁻²)	σ_y (MN m⁻²)	σ_u (MN m⁻²)	σ_b (MN m⁻²)	ϵ (%)	ρ (10^3 kg m⁻³)	ρ/σ	$\rho/\sigma_b^{0.67}$	$\rho/\sqrt{E}$	$E\rho^2/\sigma^2$
Cow femur	14(1)	150(1)	180(1)	250(1)	0.5(7)	2.1(1)	12	50	18	19
Deer antler	7(1)	—	—	180(1)	—	1.9(1)	—	62	23	—
Whale bulla	31(1)	—	—	33(1)	—	2.5(1)	—	230	14	—
Cow cartilage	6×10^{-3}(2)	—	12(2)	—	70*(2)	1.1?	92	—	450	0.5
Dog tendon	1(3)	—	100(3)	—	10(3)	1.1?	11	—	35	1.2
Cow ligamentum nuchae	5×10^{-4}(4)	—	1.7(4)	—	80*(4)	1.1?	—	—	1600	2.1
Locust stiff cuticle	10(5)	—	100(5)	—	3*(5)	1.2(5)	12	—	12	14
Locust apodeme	20(6)	—	600(6)	—	3(6)	1.1?	1.8	—	7.8	0.7
Locust resilin	2×10^{-3}(6)	—	3(6)	—	140*(6)	1.1?	370	—	780	2.7
Crab carapace	12(7)	—	32(7)	116(7)	—	1.9(7)	59	79	17	420
Spider frame silk	20(8)	1400(8)	1400(8)	—	30*(8)	1.1?	0.79	—	8	0.12
Spider viscid silk	0.55(8)	920(8)	920(8)	—	285*(8)	1.1?	1.2	—	47	0.0079
Scallop abductin	2×10^{-3}(9)	—	—	—	—	1.1?	—	—	780	—

References in brackets: (1) Currey (1979); (2) Woo, Akeson & Jemmot (1976); (3) Alexander (1974); (4) Minns, Soden & Jackson (1973); (5) Jenson & Weis-Fogh (1962); (6) Bennet-Clark (1976); (7) Wainwright, Biggs, Currey & Gosline (1976); (8) Denny (1976); (9) Alexander (1966).

and a shallow depression in the other (Fig. 5). The forces are large, but the excursions possible are small, typically only 20° or so, only one degree of freedom is used, and so the joints can be simple.

The vertebrates and the arthropods show different types of joint structure, mainly because of their use of the endoskeleton and the exoskeleton respectively, and also because of the different materials used on the bearing surfaces. The vertebrates have an endoskeleton and the sliding surfaces are covered in cartilage. Cartilage is not strong in compression, and therefore must not be exposed to high stresses. Since joints of limbs transmit large loads (Hughes, Paul & Kenedi, 1970) the area over which contact takes place must be large. This implies large radii of curvature which, if the joint excursion is large, in turn implies large areas of bearing surface on at least one of the limb bones (Fig. 6f). Usually both surfaces are large. If the forces are large, and the joint surfaces are large and have large radii of curvature, there arise great problems of stabilising the joint. It is instructive to compare the human knee and hip joints. The range of movement of the hip joint is considerable: flexion–extension 170°, abduction–adduction 75°, rotation 90°. The knee is much more restricted; flexion 160°, rotation 70°. Nevertheless, the hip joint is a fairly stable one, being a ball and socket. The head of the femur will stay in place in the acetabulum if all the ligaments and tendons round it are cut; the knee would, in similar circumstances, fall apart. The ligaments of vertebrate joints therefore function to keep the surfaces àpposed and to allow movement in certain directions and not in others. A fine example is the cruciate ligaments of the knee. These lie in the anteroposterior mid-line of the knee joint, and function both to prevent hyperextension and to determine precisely the way in which the femur slides and rolls over the tibial plateau (Frankel & Burstein, 1970). Fig. 7 is a diagram of the

Fig. 5. Cross-sections of hinge regions of various bivalve molluscs. The part of the hinge ligament loaded in compression is dotted, the part loaded in tension has fine lines.

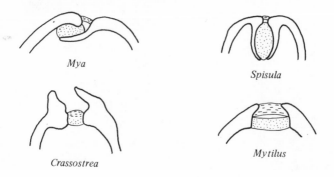

Mya

Spisula

Crassostrea

Mytilus

Fig. 6. (*a*) The material of the surface of sliding joints in arthropods is strong, therefore its cross-sectional area can be small compared with vertebrates (*d*). (*b*) The radius of curvature can be small, and so a joint region with a considerable excursion can be small (*c*). In vertebrates as the area must be large, the radius of curvature must be large (*e*) and for a similar joint excursion the joint structure must be large (*f*).

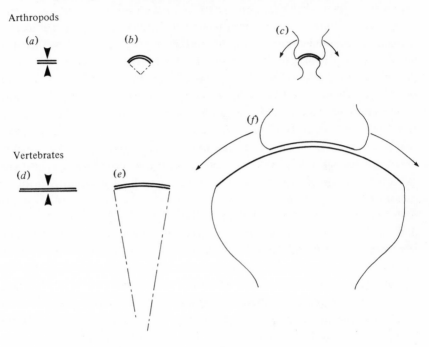

Arthropods

(*a*)　　　(*b*)　　　(*c*)

Vertebrates

(*d*)　　　(*e*)

(*f*)

Fig. 7. Front and back views of the skeleton of the wrist of the bat *Plecotus auritus*. Bones are drawn with fine lines, ligaments with thick lines and fine hachuring. After Norberg (1970).

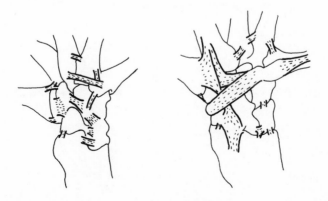

wrist joint of a bat, seen from both sides, showing the net of ligaments typically necessary on a complex vertebrate joint.

Arthropods have a generally different mode of making joints. There is a much greater tendency to have podomeres articulating with a single degree of freedom between them. Overall flexibility of movement of a limb is achieved by having a larger number of joints than in vertebrates, each with its axis at a different angle from those of its neighbours. Furthermore, the arthropods rely much more than the vertebrates on flexible hinges than on sliding joints. At the simplest, the hard cuticle is replaced over most of the perimeter of the podomere by soft arthrodial membranes (Fig. 8a). There is usually somewhere in the perimeter a straight hinge of thick but pliant cuticle (Figs. 8b, 9a). This allows fairly large forces to be transmitted across the hinge, and the pliability of the hinge material allows rotation whose magnitude is limited only by the amount that the hard cuticle is replaced by arthrodial membrane. This arrangement has the severe disadvantage, however, that it is possible to have muscles acting as flexors but not, easily, as extensors. Extension must be achieved either by the elasticity of the hinge material (as in the ischiopodite–meropodite joint of the lobster), or by a hydrostatic mechanism, that blows the limb straight, as in many spider joints. If extensors are used the hinge, and the axis of rotation, must not be on one side of the podomere but in the middle. But a hinge cannot pass right across the lumen of an exoskeleton without posing insuperable problems at ecdysis. Therefore arthropods frequently use cuticular buttresses. These extend inwards along the hinge line, but do not meet in the middle. Between the buttresses of neighbouring podomeres is a strip of pliant material (Figs. 8c,

Fig. 8. Three types of arthropod joint not using sliding surfaces. (a) Arthrodial membrane. (b) Flexible hinge. (c) Buttress joint.

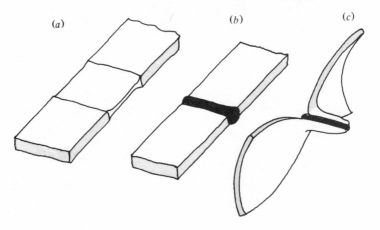

(a)   (b)   (c)

9*b*). This arrangement allows a joint capable of considerable excursion, but nevertheless also capable of bearing large forces and being actively flexed and extended by muscles.

Arthropods do also have sliding joints. For instance, the carpopodite–propodite joint of the lobster has two pairs of semicircular bearings that provide an articulation with one degree of freedom only (Fig. 10). These

Fig. 9. (*a*) Diagram of an elongate flexible hinge of an arthropod. Arthrodial membranes and soft tissues not shown. The podomere must be flattened near the hinge. (*b*) Diagram of two podomeres connected by a buttress joint.

(*a*)

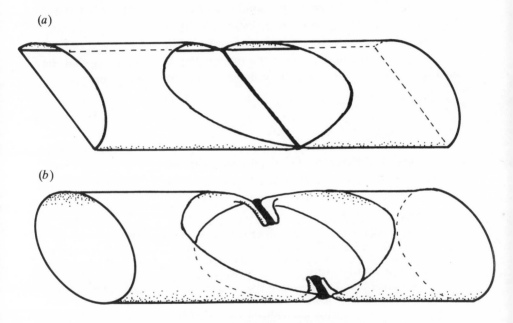

(*b*)

Fig. 10. Sagittal section of the carpopodite–propodite joint of a lobster. The hair-like extensions of the cuticle help to keep the bearing surfaces clear of grit. H – · – H is the axis of rotation of the joint.

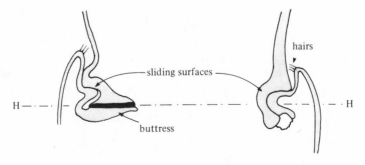

bearings are arranged so that the two podomeres can be held together by them alone: if all soft tissues and tendons are cut away the podomeres remain articulated. The joint surfaces are a very snug fit, and in a recently moulted animal are smooth and shiny. However, these beautiful bearings are exposed to the outside world, and sooner or later grit is trapped between the surfaces, which become scored. Electron micrographs show no very marked differences between these bearing surfaces and the rest of the cuticle, though they are somewhat softer.

Finally, consider as an extremely elegant design, the joint described by Darnhofer-Demar (1973) which has a very large excursion. This is the femoro–tibial joint of *Gerris*, the pond skater (Fig. 11). This has flexible hinges on buttresses. These are aided by semicircular sliding joints which are not exposed to the outer world. The tendons have intercalated hard and soft parts which act like the patella in the mammalian knee, ensuring that the muscles controlling the joint always have a large turning moment.

Tendons

The rigid elements of skeletal systems are allowed to move relative to each other at joints, and are made to do so by muscles. The muscles may connect directly with the rigid elements, or they may act via a tendon or

Fig. 11. Diagram of the femoro-tibial joint of the pond skater *Gerris najas*. Stiff cuticle drawn with thick lines, pliant cuticle with cross-hatching, arthrodial membrane with fine wavy lines. (*a*) Joint fully extended. The axis of rotation is shown by the black circle. (*b*) Joint fully flexed. (*c*) Highly diagrammatic cross-section of the joint in the plane of the hinge line. This is at right angles to the views in 11*a* and 11*b*. The bearing surfaces indicated by arrows. The pliant cuticle of the buttresses cross-hatched. After Darnhofer-Demar (1973).

(*a*)   (*b*)   (*c*)

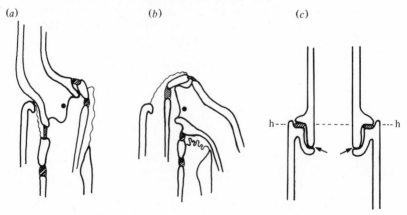

analogous structure. The functions that tendons may have are interestingly varied. I list some below.

(*a*) The presence of a tendon allows muscles to be pennate. As shown by Alexander (1968) pennation allows a muscle, under a set of not very restrictive anatomical constraints, to exert a larger force than could a parallel-fibred muscle of the same volume. This solution to the problem of exerting a large force when the cross-sectional area, but not the length, is limited, is particularly useful in arthropods, whose muscles are often contained within long narrow exoskeletal podomeres.

(*b*) A tendon allows the muscle to be far from the joint whose rotation it controls. This is useful when it would be inconvenient to have the bulk of the muscles near the joints. Such a constraint is seen in the human foot. Most of the flexors and extensors of the toes are in the lower leg, long tendons extending from there to insert on the toes. These muscles are large, and it would be cumbersome to have them in the foot itself. Having muscles positioned proximally with long tendons going distally is useful in that the main mass of the muscles can be near to the centre of rotation of the limb, so reducing the moment of inertia of the limb about its axis of rotation. This is seen particularly clearly in large cursorial animals such as the horse.

(*c*) Tendons allow the force produced by muscles to be delivered in particular places in precisely controlled directions. For instance, in the human leg the peroneus brevis has its origin on the fibula and inserts onto the base of the fifth metatarsal. It inserts via a tendon that turns through almost a right angle, being prevented from straightening by two ligamentous retinacula (Fig. 12). This arrangement would not be possible if the thick muscle had to insert directly onto the metatarsal.

Fig. 12. Diagram of the tendon of the human peroneus brevis showing how it is constrained to alter its line of action by two retinacula.

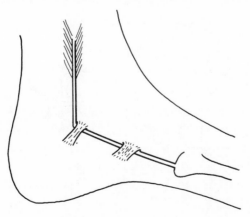

Another example of precise positioning of muscular action is in the femur–patella joints of many spiders (Fig. 13). This joint has a large range of action, about 100°. Some of the flexors do not insert directly onto the podomere, but via a crescent-shaped lump of sclerotised cuticle which is itself attached by ligaments to the podomere edge. This arrangement allows different muscle groups to have a large turning moment about the joint axis at all stages of the joint excursion. This adaptive arrangement would not be possible if all the muscles inserted directly onto the edge of the podomere; they would interfere with each other.

(d) Tendons allow for some elastic or viscoelastic deformation between the muscle and the limb elements. This may give some margin of safety, by giving the muscle time to relax and so reducing the risk of avulsion should two limb elements be forced to rotate rapidly with respect to each other. However, there is another side to the coin, insofar as if the tendon prevents the muscle from being stretched too rapidly then the muscle will be prevented from moving the joint rapidly and accurately.

(e) Tendons can act as energy stores. This is dealt with below.

What characteristics are required by a material that is to act as a tendon? The functions '*d*' and '*e*' above: acting as a safety device, require characteristics such as high compliance, that are in opposition to the other functions, so tendons can possess them only to a slight degree or be specialised for this

Fig. 13. Femur–patella joint of the spider *Ciniflo*. After Manton (1977). (*a*) Side view of the extended joint. (*b*) Side view of the flexed joint. (*c*) Ventral view of the extended joint. Arthrodial membrane not shown. Two muscles, x and y, are shown. x inserts virtually directly onto the podomere, y via a crescent-shaped intercalated lump of hard cuticle and ligaments. In (*a*) x has a small turning moment about the hinge line, while y has a large one. In (*b*) the situation is reversed. Note: there are no extensor muscles for this joint.

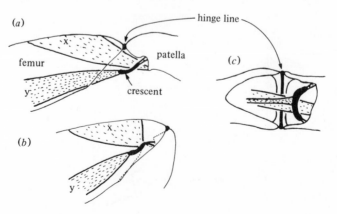

function, and in doing so will tend to be less efficient in carrying out the other functions.

The main requirements for functions a, b and c is that the ratio of Young's modulus to volume should be high. This will ensure that the mass and the volume of the tendon necessary to bear a certain load and be adequately stiff will be small. In fact, all tendon-like materials have densities in the range of 1100 to 1300 kg m^{-3}, so tendons do not differ importantly in this respect, and we can consider a high or low (property/mass) ratio as being equivalent also to a high or low (property/volume) ratio.

Vertebrate tendon and arthropod tendon are interestingly different, in that the apodemes of arthropods are about twenty times as stiff and six times as strong as tendon (Table 3). They differ little in their densities. This means, of course, that the volume of apodeme necessary to bear a particular muscular force will be much less than the amount of tendon needed. This may occasionally be of importance in arthropod podomeres where space is at a premium, being limited by the exoskeleton. This is especially so near the joints which, for reasons given above, are generally much more compact in arthropods than in vertebrates. However, the considerable stiffness of arthropod tendons makes it difficult for them to have bends with small radii of curvature as in the peroneus brevis. This problem is frequently obviated by the intercalation in the tendon of a very short segment of pliant cuticle as in *Gerris* described above (Fig. 11).

Energy stores

The ability of skeletal structures to store energy is often an important part of their function. Strain energy storage has two rather different purposes.

(*a*) One function is to amplify the power output of muscle. The muscles of small animals often have difficulty in producing work at a great enough rate to achieve the velocities required for jumping and other high-speed activities. This is best shown by the classic study of Bennet-Clark & Lucey on the flea (1967). The power output of the muscles available is grossly inadequate because the flea has to accelerate so rapidly. The flea gets over this problem by storing the necessary energy relatively slowly in elastic elements (pads of resilin) and then, by means of a catch mechanism, allowing the energy to be released suddenly. Other animals use resilin, or ordinary cuticle, or tendon to perform similar functions. The mechanism of power amplification has been worked out in some detail in, for example, the cases of the click beetle (Evans, 1973), the locust (Bennet-Clark, 1975; Heitler,

1974, 1977), the flea (Bennet-Clark & Lucey, 1967) and Mantis shrimps (Burrows, 1969).

(*b*) The other function of energy stores is to store energy in one stage of a locomotory cycle and release it at some other stage. Usually there is a change in direction of movement of an appendage or body segment, and the kinetic energy would be lost if the appendage had to be slowed down and then accelerated again by muscular action. Examples of the use of elastic storage systems are found in the resilin wing hinges of many insects, the anterior interosseus tendon of sheep (Cuming, Alexander & Jayes, 1978), the ligamentum nuchae of many large mammals (Alexander, 1968), the gastrocnemius tendon of kangaroos (Alexander & Vernon, 1975), the skin of sharks (Wainwright *et al.*, 1978) and the hinge ligaments of scallops and of burrowing bivalve molluscs (Trueman, 1975).

The characteristics required of elastic storage elements depend on their particular function. However, the two most important are probably their (strain energy)/mass ratio, and their resilience.

The values for the maximum elastic energy-absorbing properties of various possible candidates for energy stores are shown in Fig. 14. The abscissa shows the maximum energy absorbed per kilogram, the ordinate the stress to which the material must be raised to achieve the energy storage. Two points are clear. First, as an energy store, muscle is orders of magnitude

Fig. 14. Diagram of energy-storage ability and stress of various tissues. Abscissa: the energy that can be absorbed by the material before it breaks. Ordinate: stress at maximum energy absorption. The proportion of the absorbed energy released is variable. It is very low in cartilage.

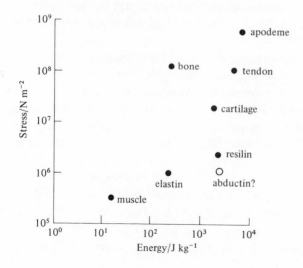

less effective than tendon. This was convincingly demonstrated by Alexander & Bennet-Clark (1977). The way in which energy is stored in locomotion by tendon has been shown in many papers by Alexander and his co-workers (Alexander & Vernon, 1975; Cuming *et al.*, 1978). A most interesting case is the anterior interosseus muscle of sheep in which the muscular portion has been reduced to a vestige, and the tendon is effectively an elastic strap running between the metacarpals and the phalanges.

The second point made by Fig. 14 is that the various high-energy-storing materials have great differences in stress implying, of course, great differences in strain. The apodeme of the locust extensor tibiae is used to store energy prior to the jump, as are the resilin pads in the coxatrochanter region of the flea. Apodeme has a low strain, of the order of 3%, and a high stress. The strain in the resilin pad is far greater, and as it stores nearly one-third of the energy per unit volume as does the apodeme, the stress is far less. I do not understand why there should be these differences in the mechanical properties of materials that function as energy stores, but here are two suggestions.

One advantage of the low-stress system is that as the stresses in the elastic material are low, the forces it exerts on the surrounding tissues are low, and so the surrounding tissues do not need to be particularly thick to withstand the loads. As Alexander (1968) showed, pennate muscle fibres produce large forces with little overall deformation of the muscle. The extensor tibiae of the locust is necessarily highly pennate, being constrained by the femoral cuticle surrounding it (Uvarov, 1966). Therefore, it is advantageous for this muscle to store energy in a high-stress small-deformation material. The trochanter depressor of the flea, on the other hand, is contained in the more spacious thorax, and is much more parallel-fibred than the locust extensor

Fig. 15. Diagram of loading–unloading curve.

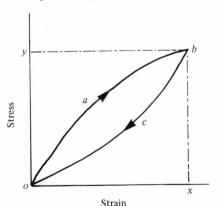

tibiae (Rothschild & Schlein, 1975). It is, therefore, better adapted to producing a relatively small force with a large contraction. Resilin is a more suitable energy-storing material for it to work against. Although virtually no work has been done on this topic, which would repay study, it may be found that the use of high-strain or low-strain stores will be largely determined by such geometrical constraints.

Another important aspect of elastic elements is their time-dependent behaviour. Suppose a material has a loading–unloading stress–strain curve, at the frequency appropriate for its function, as in Fig. 15. On being loaded to stress y, and strain x, the material will absorb energy proportional to the area $oabx$. On unloading an amount of strain energy proportional to $ocbx$ will be available for doing mechanical work. The ratio $ocbx/oabx$ is the 'resilience', the greater it is the less energy lost, as heat, in the loading–unloading cycle.

If a tendon is stretched and held at a particular length, the load required falls with time. This viscoelastic behaviour is caused by the internal rearrangement of the molecules of the tendon, which tend to occupy a position of minimum strain energy. Therefore, the amount of recoverable energy stored in the tendon decreases with time (Cuming et al., 1978). Resilin, on the other hand, is remarkably good at retaining its energy over long times, a property not made much use of in locomotion, when the cycle time is short. It is useful in some situations, however. In dragonflies, for instance, a muscle contracts to keep a resilin tendon at a certain average length, the tendon itself automatically modifying the twist of the wing (Neville, 1965).

The hinge ligaments of bivalves are used to force the valves apart when the adductors relax. In burrowing bivalves this is a vital part of the locomotory cycle (Trueman, 1954). It is important that the ligament should not relax greatly, as there is no other means of producing the necessary force. Little seems to be known about stress relaxation (as opposed to short-term hysteresis) in hinge ligaments in burrowing bivalves. Kahler, Fisher & Sass (1976) examined the hysteresis of eight bivalve species' hinge ligaments. The loading–unloading frequency was not reported, but was probably about 10^{-2} Hz. The resilience ranged from 82% in Ensis, the razorshell, to 97% (a remarkably high value) in Placopecten, the smooth scallop. Alexander (1966) reported that the abductin hinge of scallops crept considerably for a few seconds, and then stopped deforming. I have measured stress relaxation in Cyprina islandica. There is a rapid fall in the force exerted by the hinge ligament, until in about two minutes it falls to 60% of the initial value. Thereafter the force declines extremely slowly, and is almost unchanged after several hours.

References

Alexander, R. McN. (1966). Rubber-like properties of the inner hinge-ligament of Pectinidae. *J. exp. Biol.* **44**, 119–30.

Alexander, R. McN. (1968). *Animal Mechanics.* London: Sidgwick & Jackson.

Alexander, R. McN. (1974). The mechanics of jumping by a dog (*Canis familiaris*). *J. Zool., Lond.* **173**, 549–73.

Alexander, R. McN. (1975). *Biomechanics.* London: Chapman & Hall.

Alexander, R. McN. & Bennet-Clark, H. C. (1977). Storage of elastic strain energy in muscles and other tissues. *Nature, Lond.* **265**, 114–17.

Alexander, R. McN. & Vernon, A. (1975). The mechanics of hopping by kangaroos. (Macropodidae). *J. Zool., Lond.* **177**, 265–303.

Bennet-Clark, H. C. (1975). The energetics of the jump of the locust. *Schistocerca gregaria. J. exp. Biol.* **63**, 53–83.

Bennet-Clark, H. C. (1976). Energy storage in jumping animals. In *Perspectives in Experimental Biology*, ed. P. Spencer Davies, *Volume 1 Zoology*, pp. 467–79. Oxford: Pergamon Press.

Bennet-Clark, H. C. & Lucey, E. C. A. (1967). The jump of the flea; a study of the energetics and a model of the mechanism. *J. exp. Biol.* **47**, 59–76.

Burrows, M. (1969). The mechanics and neural control of the prey capture strike in the mantid shrimps *Squilla* and *Hemisquilla. Z. vgl. Physiol.* **62**, 361–81.

Cavagna, G. A., Saibene, F. P. & Margaria, R. (1964). Mechanical work in running. *J. appl. Physiol.* **19**, 249–56.

Cuming, W. G., Alexander, R. McN. & Jayes, A. S. (1978). Rebound resilience of tendons in the feet of sheep (*Ovis aries*). *J. exp. Biol.* **74**, 75–81.

Currey, J. D. (1967). The failure of exoskeletons and endoskeletons. *J. Morphol.* **123**, 1–16.

Currey, J. D. (1979). Mechanical properties of bone tissues with greatly differing functions. *J. Biomechanics,* **12**, 313–19.

Darnhofer-Demar, B. (1973). Funktionsmorphologie eines Arthropodengelenks mit extrem grossen Aktionswinkel: Das Femoro-Tibial-Gelenk von *Gerris najas* (De G.) (Insecta, Heteroptera). *Z. Morphol. Tiere,* **75**, 111–24.

Denny, M. (1976). The physical properties of spider's silk and their role in the design of orbwebs. *J. exp. Biol.* **65**, 483–506.

Evans, M. E. G. (1973). The jump of the click beetle (Coleoptera: Elateridae) – energetics and mechanics. *J. Zool., Lond.* **169**, 181–94.

Frankel, V. H. & Burstein, A. H. (1970). *Orthopaedic Biomechanics.* Philadelphia: Lea & Febiger.

Headley, F. W. (1895). *The Structure and Life of Birds.* London: Macmillan.

Heitler, W. J. (1974). The locust jump: specialisations of the metathoracic femoral–tibial joint. *J. comp. Physiol.* **89**, 93–104.

Heitler, W. J. (1977). The locust jump III. Structural specialisations of the metathoracic tibiae. *J. exp. Biol.* **67**, 29–36.

Hughes, J., Paul, J. P. & Kenedi, R. M. (1970). Control and movement of the lower limbs. In *Modern Trends in Biomechanics*, vol. 1, pp. 147–79, ed. D. C. Simpson. London: Butterworths.

Jensen, M. & Weis-Fogh, T. (1962). Biology and physics of locust flight, v. strength and elasticity of locust cuticle. *Phil. Trans. R. Soc., Ser. B.*, **245**, 137–69.

Kahler, G. A., Fisher, F. M. & Sass, R. L. (1976). The chemical composition and mechanical properties of the hinge ligament in bivalve molluscs. *Biol. Bull.* **151**, 161–81.

McMahon, T. A. & Kronauer, R. E. (1976). Tree structures: deducing the principle of mechanical design. *J. theor. Biol.* **59**, 443–66.

Manton, S. M. (1977). *The Arthropoda, Habits, Functional Morphology, and Evolution*. Oxford University Press.

Minns, R. J., Soden, P. D. & Jackson, D. S. (1973). The role of the fibrous components and ground substance in the mechanical properties of biological tissues: a preliminary investigation. *J. Biomechanics*, **6**, 153–65.

Neville, A. C. (1965). Energy and economy in insect flight. *Sci. Progr. Oxf.* **53**, 203–19.

Norberg, U. M. (1970). Functional osteology and myology of the wing of *Plecotus auritus* Linnaeus (chiroptera). *Ark. Zool.* **22**, 483–543.

Rothschild, M. & Schlein, J. (1975). The jumping mechanism of *Xenopsylla cheopis*. 1. Exoskeletal structures and musculature. *Phil. Trans. R. Soc., Ser. B.*, **271**, 457–90.

Shanley, F. R. (1960). *Weight-Strength Analysis of Aircraft Structures*. New York: Dover.

Trueman, E. R. (1954). Observations on the mechanism of the opening of the valves of a burrowing lamellibranch, *Mya arenaria*. *J. exp. Biol.* **31**, 291–305.

Trueman, E. R. (1975). *The Locomotion of Soft-bodied Animals*. London: Edward Arnold.

Uvarov, B. (1966). *Grasshoppers and Locusts*, vol. 1. Cambridge University Press.

Vincent, J. F. V. (1975). Locust oviposition: stress softening of the extensible intersegmental membranes. *Proc. R. Soc., Ser. B*, **188**, 189–201.

Wainwright, S. A., Biggs, W. D., Currey, J. D. & Gosline, J. M. (1976). Mechanical design in organisms. London: Edward Arnold.

Wainwright, S. A., Vosburgh, F. & Hebrank, J. H. (1978). Shark skin: function in locomotion. *Science, NY*, **202**, 747–9.

Woo, S. L-Y., Akeson, W. H. & Jemmott, G. F. (1976). Measurements of nonhomogeneous directional mechanical properties of articular cartilage in tension. *J. Biomechanics*, **9**, 785–91.

M. A. SLEIGH & D. I. BARLOW

Metachronism and control of locomotion in animals with many propulsive structures

There has been a tendency during evolution for a reduction in the number of limbs or other propulsive structures that are used in locomotion. The majority of terrestrial vertebrates and arthropods walk, run or jump using one, two, three or four pairs of legs, and their swimming relatives use a similar number of paired appendages. However, among the invertebrates are many groups in which large numbers of propulsive structures – limbs, tube feet or cilia – take part in coordinated locomotion, either swimming or walking. In this chapter the present state of knowledge concerning this coordinated locomotion will be reviewed.

The requirements and basic features of control and coordination

Structures used for propulsion clearly need to be under the control of the organism so that it can regulate whether it shall move, how fast it shall move and the direction in which it shall move. All organisms seem therefore to have developed a mechanism which determines whether movement shall take place or not, some organisms depending upon activation and others upon inhibition to achieve this control. In most cases the rate of cyclical activity of propulsive structures is regulated, the control mechanism involving activation in some organisms and a combination of activation and inhibition in other cases. Control of the direction of movement depends upon the form of propulsive structure that is used. The locomotive tube feet of many echinoderms and the cilia in some organisms where these are used for locomotion are capable of propulsion in many different compass directions relative to the surface which bears the propulsive structures; control systems exist which coordinate changes in direction so that the compass direction of movement is consistent over the whole body. Certain paddle-shaped structures, notably the compound cilia of ctenophores and the parapodia of annelids, can execute either forward power strokes or reverse power strokes, and the synchronous reversal of activity indicates a centralised control system. Finally, the limbs of other animals, e.g. some arthropods, are not capable of reversed propulsive movement, and alternative means are used to direct locomotion, e.g. steering or tail-flip movements of the abdomen or telson of a swimming crustacean, steering by body bending of a myriapod, or the thrust direction can be vectored by the limbs in some Branchiopoda.

In most cases where numerous propulsive structures are used together there is also a well-established rhythm of activity of the component structures, so that adjacent members of the series maintain a constant phase relationship, and metachronal waves of activity can be seen to pass over the area of ciliated surface or along the rows of limbs, parapodia or cilia. The maintenance of such a metachronal rhythm requires both a mechanism for generating regular excitation of the component structures of the system and a mechanism for communicating information about the phase of cyclical activity between adjacent members of the system.

The form of metachrony is different in different cases. It is described by relating the direction of movement of the metachronal waves with the direction of propulsion. However, different terminology has developed for ciliated systems and limb systems; because many ciliated systems propel fluids over stationary surfaces, Knight-Jones (1954) defined ciliary metachronal patterns in terms of the relationship of the direction of the propulsive stroke of the cilia to the direction of propagation of the metachronal wave, but it is conventional when discussing limb systems to relate the direction of metachronal waves to the direction of movement of the organisms. Thus in those metachronal systems where the waves of coordination travel from the anterior of the moving animal towards its posterior the waves are referred to as symplectic when they involve cilia, because waves and power strokes are in the same direction, or as retrograde or contralocomotory waves when they involve limbs. Where metachronal waves travel from the posterior end towards the anterior end of a moving animal, ciliary metachrony involves antiplectic waves, because waves and power strokes are in opposite directions, whilst limb metachrony is described as direct or adlocomotory. In many ciliary systems, but seldom if ever in limb systems, the direction of propulsion is at right angles to the direction of propagation of metachronal waves; the metachrony here is diaplectic, and may be either dexioplectic, when the power strokes take place to the right when viewed in the direction that waves are being propagated, or laeoplectic, when the power strokes of the cilia move towards the left when the cilia are viewed so that the waves are travelling away from the observer.

Whether the component propulsive structures of these systems move in metachronal rhythm or not, they tend to show cyclical activity of power strokes and recovery phases. There is wide agreement that each limb, cilium or tube foot has the potentiality of generating a rhythmical cycle of activity, and contains or is associated with some 'oscillator' mechanism. This oscillator provides the instructions for the timing of events within the cycle of activities, but the frequency and often the phase of the oscillator can be influenced by the organism or by the activity of neighbouring structures. In some cases the form of the mechanical output of the structure, but often not the oscillator activity, may be modified by signals from a direction-controlling system in the animal. In multicellular animals the control systems, some of the coordination systems, and often the oscillator itself, appear to be functions of the nervous system, but in unicellular forms simpler control and regulator systems have been developed.

The tube feet of echinoderms

The use of the tube feet by echinoderms for locomotion has been most thoroughly studied in asteroid starfish, particularly by Smith (1950, 1966). Each of the five arms of the regular asteroid has two rows of tube feet running the length of the oral (ambulacral) surface, one row at either side of the radial nerve in the ambulacral groove. These tube feet bear terminal suckers which can attach to a substratum and provide locomotory forces; complex systems of intrinsic (and ampullary) muscles cause (a) the extension of the tube foot in a particular direction, (b) the attachment of the sucker to the substratum, (c) a swinging motion of the tube foot to move the body relative to the sucker attachment, (d) the release of the sucker, and finally, (e) a retraction of the tube foot.

Smith observed that each tube foot of a moving starfish tended to show a regular stepping rhythm with cyclical extension and retraction. The tube feet do not show any regular phase relationship in these cycles of activity among the members of a row along an arm or between members of the two rows at either side of the radial nerve. Changes in the speed of movement are achieved by changes in pace of the cyclical stepping rhythm, and all tube feet of an animal change their pace simultaneously. Directed locomotion is achieved because the compass-direction of all tube feet of all of the arms has a common alignment, and when the direction of locomotion changes, the alignment of the stepping cycle of all tube feet changes together.

The actions of protraction and retraction of the tube foot are thought to be under peripheral control and reflect oscillatory conditions of excitation and inhibition of neurons controlling the ampulla and retractor muscles of the foot. Although Smith (1950) found that external stimuli caused retraction and protraction, he tentatively suggested that retraction might be stimulated by a high-frequency burst of impulses reaching the terminal motor neurons of the retractor muscles whilst a low-frequency after-discharge stimulated the ampulla motor neurons to cause protraction, the whole sequence being set up by an oscillator system possibly located within the radial nerve, but not under central control. External stimuli would modify the significance of the high- and low-frequency portions of the signal to cause a bias towards retraction or protraction depending on the type of external stimulus received.

In contrast to the peripheral control of stepping, the pivoting response of the feet to step in any particular direction appears to be under the control of the central pathways. The pivoting response is controlled by states of excitation and inhibition in antagonistic pairs of neurons stimulating postural muscles arranged around the tube foot. It is difficult to reconcile the two independent sources of stimulation which must interact in a set sequence if proper locomotion is to result.

Fig. 1. A schematic hypothesis of the control of tube feet of asteroid echinoderms. An oscillator circuit is associated with the group of motor neurons (M) concerned with the protraction/retraction and postural control of each tube foot. Rate control is envisaged as modifying the frequency of the oscillator circuit, while directional control is supposed to influence the postural control neurons (P) associated with the motor centre.

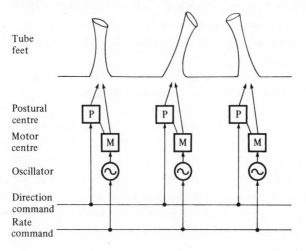

Smith noted that an arm severed from the circum-oral ring would walk towards the base whilst an isolated arm still attached to a piece of circum-oral ring would walk towards the tip. Also, if the circum-oral ring was cut between each arm then all five arms would tend to walk towards their tips. Each arm is assumed to possess a direction control centre, probably situated at the junction of the radial nerve with the circum-oral ring; in the intact animal one direction control centre dominates the rest and the arm associated with this centre becomes the leading one. A decline in the activity of one centre is matched by an increased excitation in an adjacent one. There appears to be a periodicity in the rise and fall of activity of the direction control centres.

In summary, the pattern of behaviour is consistent with the suggestion put forward by Smith (1966) that each tube foot has its own set of motor neurons which generate its stepping rhythm, and that the level of activity and direction of stepping may be imposed on the individual neurons of each tube foot by other parts of the nervous system. A diagram illustrating the basic features of oscillator, rate command and direction command components of this system is given in Fig. 1.

Echinoderm tube feet provide the only example we have found in which multiple propulsive structures do not move in metachronal rhythm. One must conclude that the benefits other systems derive from metachrony do not apply to tube feet. If metachrony of tube feet does occur, it might be expected to be found in forms that burrow in soft substrates.

Cilia

The cilia of a group normally have the same pattern and frequency of beating but seldom have the same phase of beating, since they normally display metachronal coordination of activity. The rate of beating of all the cilia of a group is under the control of the organism, at least to the extent of being active or inactive, and in many organisms there is a finer control of frequency. The direction of the power stroke of the beat is also variable in some cases, under central control by the organism, so that the direction of beat of a whole group of cilia changes in a coordinated fashion.

The rhythmic beat of cilia and flagella seems to be an intrinsic property of the mechanism associated with the fibril bundle of the axoneme. Naked ciliary or flagellar axonemes, produced by detachment of the shaft distal to the basal body and removal of the covering membrane with detergent, can undergo repeated rhythmic undulations in the presence of ATP and appropriate concentrations of ions (see review by Goldstein, 1974). The ciliary

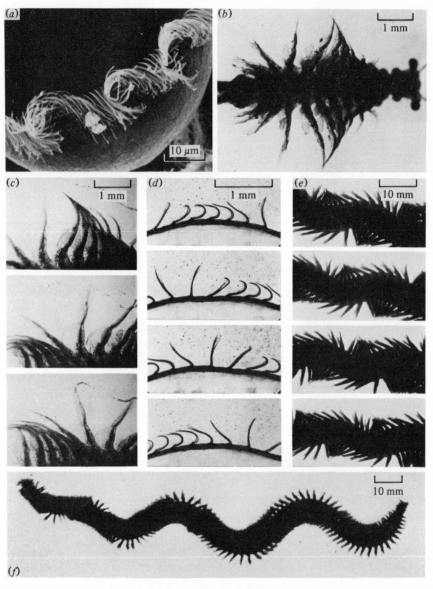

Fig. 2. (*a*) Metachronism of the swimming cilia of *Didinium nasutum*. (Scanning electron micrograph by technique of Barlow & Sleigh (1979).) (*b*) Metachronism in *Artemia salina* – the limbs on both sides of the same segment move in phase. (*c*) Three frames from a high-speed film of *Artemia* to show limbs of one side at 0, 130 and 260 ms, one cycle taking 330 ms. (*d*) Metachronism of the comb plates of *Pleurobrachia pileus* shown in four high-speed frames at 0, 13, 20 and 30 ms, one cycle taking 40 ms. (*e*) Metachronism of parapodia of a walking *Nereis diversicolor*; four movie frames show positions of parapodia at 0, 313, 626 and

'oscillator' is therefore believed to be within the axoneme, and to involve repeated alternations of motion of components of a mechanochemical mechanism formed by the longitudinal fibre bundle and associated transverse elements.

Cilia or flagella normally beat in coordinated waves. When many spermatozoa lie side by side with similar orientation, they tend to share the same rhythm, the bending waves of adjacent sperm tails being synchronised with one another (Gray, 1928). The independent symbiotic spirochaetes attached to the surface of the flagellate protozoan *Mixotricha* show metachronally coordinated undulations; the spirochaetes are attached obliquely to the surface, and although adjacent parts of neighbouring spirochaetes move in synchrony the moving waves involve the population of spirochaetes in a metachronal sequence (Cleveland & Cleveland, 1966). More conventional metachronal waves of similar appearance occur on such multiflagellate Protozoa as *Deltotrichonympha* (Cleveland & Cleveland, 1966) or on *Opalina* (Sleigh, 1960; Tamm & Horridge, 1970). The metachronal waves of these systems are symplectic. In the majority of ciliated Protozoa the metachronism of the cilia is dexioplectic, both in cases where the cilia cover an extensive surface, e.g. *Paramecium* (Machemer, 1972), and where there are narrow bands of cilia, e.g. *Stentor* (Sleigh, 1956) or *Didinium* (Parducz, 1961) (Fig. 2*a*). Locomotor cilia using laeoplectic waves occur in molluscs, e.g. veliger larvae (Sleigh, 1969), while dexioplectic metachronal waves are usual on rotifers, and such larvae as those of echinoderms and annelids (Knight-Jones, 1954). An unusual form of antiplectic metachronism is shown by the comb plates of ctenophores, which are giant paddle-shaped compound cilia (Sleigh, 1968) (Fig. 2*d*).

The coordination of cilia is achieved by the entrainment of the intrinsic rhythmical activity of the component cilia, using a timing signal communicated between neighbouring cilia of the group. Normally, and perhaps always, this signal takes the form of viscomechanical coupling through the intervening water (Machemer, 1974). The synchronisation of adjacent flagella or the metachronism of *Opalina* cilia provide examples in which each of the close-set organelles generates viscomechanical forces which influence the movement of all immediately adjacent organelles; this gives rise to

---

939 ms, one cycle taking about 1252 ms. Propulsive strokes of the limbs take place from right to left whilst the body undulations can be seen to move from left to right and the body is travelling towards the right. (*f*) Metachronism of parapodia of a swimming *Nereis diversicolor*.

continuous mechanical interference which can provide the necessary timing information to entrain the axonemal oscillators of the cilia or flagella involved. The cilia which take part in diaplectic metachronal patterns show non-planar beating cycles in which the cilia move to one side or the other during the recovery phase; it is found that the metachronal waves are always propagated in the direction towards which the cilia move in their recovery phase. Thus the dexioplectic metachrony of *Paramecium* is associated with a counter-clockwise motion of the cilium, as seen from the tip (Machemer, 1972), whilst the laeoplectic metachrony of molluscs is associated with a clockwise movement of the ciliary tip during the recovery phase, as seen when looking from tip to base (Aiello & Sleigh, 1972). In these cases the viscomechanical interaction is probably not effective throughout the cycle, but is thought to provide sufficient coupling in the recovery stroke to determine the relative phase of the beat cycles of adjacent cilia in the rows along which metachronal waves pass. Where the rows are more than one cilium wide, the cilia at right angles to the row (and in the plane of the power strokes) beat almost synchronously because they are strongly coupled by viscomechanical forces acting in the power strokes of the beat.

The antiplectic metachrony of ctenophore comb plates appears to provide a special case in two respects, first that the cilia seem to have a low level of intrinsic excitability, so that they lie still unless stimulated, second that each beat of a comb plate is normally triggered individually by the activity of the next comb plate of the row (Sleigh, 1972, 1974). Any individual comb plate possesses the ability to generate normal repetitive cycles of rhythmic beating, and will do so if sufficiently excited, e.g. by local damage, but normally this intrinsic rhythmicity is suppressed. In non-lobate ctenophores like *Pleurobrachia* and *Beroë* the movement of one comb plate lifts the plate that lies below it, and in doing so stimulates it to begin its active stroke and in turn to trigger the next plate below (Sleigh, 1972). Each wave of normal beating in a comb row begins with a beat by the top plate in the row; this itself is stimulated, in a manner that has not been resolved, by a wave of beating which passes along a band of cilia (ciliated groove) from the balancer cilia of the apical sense organ. This structure is connected to each of the eight comb rows and normally regulates the activity of every comb plate of the body, a beat of any balancer cilium being followed by a wave of activity and a single beat of each of the comb plates in the two associated comb rows (Fig. 3) (Horridge, 1965). Slower symplectic waves can be conducted up the comb rows under abnormal conditions (Sleigh, 1968). In the lobate ctenophores the triggering of one plate by the plate above can take place in the absence of the strong hydromechanical coupling seen in *Pleurobrachia*; in such lobate forms as *Mnemiopsis* segments of the ciliated grooves are found between each plate throughout the length of the comb row, and these may be involved in the triggering action (Tamm, 1973).

In no case is it known how the organism controls the frequency of the ciliary oscillator, although it is suspected that the action on the axonemal mechanism is through ionic levels or possibly energy supply. External mechanical restraints (e.g. changes in viscosity) can affect the frequency of beating and internal control could be exerted through changes in internal resistance to sliding of the axonemal microtubules, or through changes in the biochemical reactions associated with the generation of active sliding forces. In detergent-extracted 'model' cilia the rate of beat varies with both

Fig. 3. The arrangement and movement of comb plates of the ctenophore *Pleurobrachia*, and their relation with other body structures. In (*a*) the arrangement of comb plates (CP) in eight rows is shown in relation to the mouth (M), ciliated grooves (CG) and apical organ (AO). (*b*) shows waves of beating that pass down a comb row, the plates performing upward effective strokes. In (*c*) the ciliated grooves pass from the top comb plate of the row to the balancer cilia (B) that support the statolith (S) in the apical organ. The comb plates in this species are about 0.5 mm long. (*a*) and (*c*) are from Sleigh (1972) and (*b*) from Sleigh (1968), with permission.

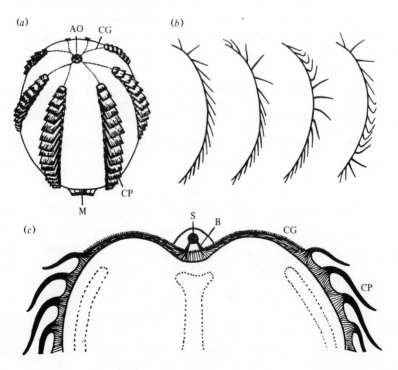

magnesium ion and ATP concentrations (Naitoh & Kaneko, 1973). All known examples involve control systems that quickly influence all the cilia of a group, either by nerves or by membrane potential changes of the ciliated cells, or both.

Frequency control of at least some mollusc cilia is exerted through the nervous system. Nerves that release 5–hydroxytryptamine cause acceleration and those that release dopamine cause a reduction in frequency. Brief inhibition of beat occurs when the gill nerve is stimulated, probably by a direct electrical effect on the ciliated cells (Aiello, 1974). The beat frequency of swimming cilia in Protozoa varies with the membrane potential of the cell, and control is therefore presumably exerted through ionic levels. Thus hyperpolarisation, which is thought to lower internal potassium ion levels, increases the frequency of beat in *Opalina*, *Paramecium* or *Euplotes* (Naitoh & Eckert, 1974); small depolarisations decrease the frequency of beat.

Where appropriate studies have been made, ciliary reversal has been found to accompany large depolarisations of the ciliated cells, either affecting the whole organism almost simultaneously in Protozoa or communicated quickly to the ciliated cells by nerves in multicellular forms. Reverse beating of locomotor cilia is only known in Metazoa among ctenophores (Barlow & Sleigh, unpublished data), but reverse beating of pharynx cilia occurs in appendicularians where it is believed that depolarisation of the ciliated cell accompanies the reversal of beat (Galt & Mackie, 1971; Mackie *et al.*, 1974). Calcium ions seem in all cases to act as the messengers which connect

Fig. 4. A schematic hypothesis of the control and coordination of cilia. The mechanochemical oscillations of elements within the ciliary axoneme are subject to rate control and often direction control, usually on a whole-cell basis by electrical/ionic changes at the cell surface. Viscomechanical forces acting in the fluid between the ciliary shafts provide phase coupling.

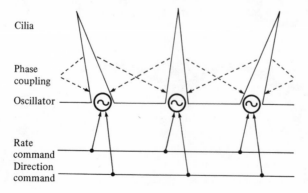

the membrane depolarisation to axonemal components that proceed to perform a reversed pattern of sliding.

The overall activity of a group of cilia is therefore believed to depend upon independent mechanochemical oscillators in each component cilium, whose frequency and phase is determined by entrainment involving viscomechanical coupling between neighbouring cilia, minimum mutual interference resulting in an average frequency and ciliary metachronism. The general level of activity of the ciliary oscillators can be raised or lowered under the control of the organism, through nerves, hormones or propagated membrane potential changes, and in many cases the direction of beating of the cilia can be changed by depolarisation of the ciliated cells under the control of the organism through nerves or direct effects on the cell membrane. A diagram illustrating the general relationships of oscillator, phase determination, rate command and direction command components is given in Fig. 4.

Appendages of arthropods and onychophorans

Onychophorans and arthropods from a number of classes possess multiple segmental appendages that move in metachronal rhythm for swimming, walking or respiratory ventilation. The more complex patterns of limb coordination found in representatives with four pairs of walking legs, or less, will not be discussed; extensive information is available elsewhere on these forms (see reviews by Delcomyn, 1977, and Dorsett, this volume). Among aquatic crustaceans the branchiopods, notably the anostracans, e.g. *Artemia* and *Chirocephalus* show clear metachrony of numerous swimming limbs; shorter series of metachronally coordinated swimming limbs are found in other crustacean subclasses, including copepods and various malacostracan groups, within which the abdominal swimmerets of several decapods, notably *Procambarus*, have been the subject of detailed research. Metachronal use of walking limbs also occurs, e.g. in the isopod *Ligia*. Insects seldom have long series of segmental appendages that move metachronally (except, perhaps, in a peristaltic mode, see Trueman & Jones, 1977), but the abdominal gills of ephemeropteran nymphs provide good examples. The onychophorans and the myriapod groups possess long series of walking legs which are moved in metachronal patterns of a variety of types.

In the crustacean examples the two limbs of a segment normally move synchronously and waves of activity pass forwards from the most posterior limb of the series in adlocomotory rhythm (Fig. 2*b, c*) (Cannon, 1933; Lowndes, 1933). In the mayflies the rhythmic beating of the nymphal gills begins with the anterior gill and moves backwards in the same direction as the propulsion of water, so that the rhythm is 'contralocomotory', although not actually concerned with locomotion of the nymph. Usually the two gills of a segment beat together and propel water backwards; in *Caenis* the gill at one side beats a little before the other and water is propelled across the body; but the sequence is not immutable and coordinated changes in the direction of water flow occur (Eastham, 1934; 1936).

Synchrony of the two limbs of a segment and adlocomotory coordination of walking limbs is usual in Diplopoda, whilst in such chilopods as *Lithobius* the adlocomotory metachronism is associated with alternate stepping of the two limbs of a segment, much as in *Nereis* (Manton, 1953; Gray, 1968). The chilopod *Cryptops* shows contralocomotory metachronism of its limbs, the two limbs of a segment again being 180° out of phase (Manton, 1953). In *Peripatus*, the two limbs of a segment are in phase during slow movements, but in antiphase for fast movements; the metachronal waves are always adlocomotory (Manton, 1950). The relative advantages of adlocomotory and contralocomotory waves for different myriapods have been discussed by Manton (1953), Knight-Jones & Macfadyen (1959) and Gray (1968).

Reversal of the direction of the power stroke of the appendage or of the progression of the wave does not seem to occur in arthropods (see below). The limb movements are, however, controlled in a variety of respects by the animal; these include changes in the phase difference between the movements of adjacent ipsilateral limbs, and changes in the form of the limb cycle by alteration of the relative duration or amplitude of power and recovery strokes or of the shape or inclination of the limb at different parts of the cycle. The most common effects of increased excitation on the activity of branchiopod limbs appear to be an increase in the frequency of beating coupled with a decrease in phase difference between adjacent pairs of limbs.

Studies of the swimmeret rhythm of *Procambarus* (see Hughes & Wiersma, 1960; Stein, 1974) have revealed interesting features of metachronal control. The five abdominal segments that bear swimmerets are innervated by nerves from segmental ganglia (composed of lateral hemiganglia) linked by ventral connectives of the nerve cord. The first nerve to arise from the ganglion in each segment contains motor nerves to the swimmerets. If the connectives in front of and behind a ganglion were cut, rhythmical motor discharge could be recorded in the swimmeret nerve, and this continued if sensory input was removed by cutting the nerves; generation of the rhythm of limb beating must occur within the central nervous tissue of each segment (Stein). In fact, since the frequency of beating of the swimmerets at either side of a segment has occasionally been observed to be different (Hughes & Wiersma), there must be a separate generator of limb rhythm for each swimmeret, i.e. in each hemi-ganglion of swimmeret-bearing segments. Stein observed rhythmic discharge in nerves running forward in the connective between one ganglion and the next. The rhythmic discharge occurred in synchrony with a motor discharge in the swimmeret nerve of the next posterior ganglion (but not synchronous with activity in the next anterior segment); the discharge in the connective also persisted in the absence of sensory input. When the connective was cut, the phase relationship between the limbs at either side of the cut was no longer maintained. In some elegant experiments Stein showed how it was possible for changes in the frequency of motor discharge in one ganglion to advance or retard the motor output of the ganglion in front and thereby regulate both frequency and phase, while there was little or no influence of a ganglion on the one behind.

The pattern of swimmeret beating in *Procambarus*, analysed largely by Stein (1974), appears therefore to depend upon three main components: (a) a system of oscillator neurons within each hemi-ganglion of each segment which generates the patterned motor output to drive the power and recovery strokes (i.e. distributed generators of locomotor rhythm entirely within the central nervous system); (b) coordination neurons which carry timing signals from one oscillator to the next anterior ipsilateral oscillator,

Fig. 5. A schematic hypothesis of the control and coordination of the swimmerets of a crayfish or the thoracic limbs of a branchiopod crustacean. The frequency of the oscillator circuit is influenced by the rate command system and the phase of its cycling is influenced by the coordinating neurons. The output of the oscillatory circuit drives the motor circuits (M) which produce a patterned response.

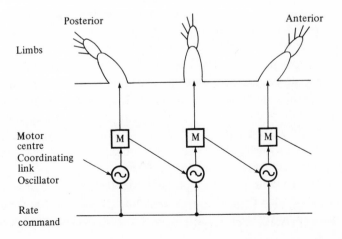

serving to advance or delay the cycle of the second oscillator, according to the time of arrival within the cycle of the modulated oscillator of the impulse from the coordinating neuron; (c) command neurons whose frequency of discharge sets the general excitation level of the oscillatory neurons – when the command neurons are stimulated at above a certain frequency the swimmerets beat more quickly. The supposed relationship of these major components is indicated in Fig. 5. Minor components presumably include coordinating elements that normally maintain contralateral synchrony, coordinating elements passing information in the posterior direction and additional command neurons able to provide instructions about changes in beat amplitude, etc. Such evidence as there is about other metachronally coordinated arthropod systems suggests that the three main components persist in other cases, although the connections made by the coordinating neurons will vary according to the pattern of coordination.

The parapodia of polychaete worms

Walking and swimming movements of many 'errant' polychaete worms involve active propulsive strokes of paddle-shaped parapodia (Fig. 2e), supplemented in faster locomotion by lateral body undulations caused by contraction of longitudinal muscles (Fig. 2f). Each main segment of the body of one of these annelids carries a lateral fleshy parapodium, supported by bundles of large setae, at either side. A double ventral nerve cord connects segmental ganglia: lateral segmental nerves arise from the ganglia, and a subsidiary parapodial ganglion may be present at each side on the nerve that supplies the parapodium. During forward walking or swimming metachronal waves of parapodial activity pass forward from tail to head; reversal of movement is associated with reversal of the metachronal waves, which then pass from head to tail.

Polychaetes that are walking slowly may use their parapodia only and keep the body straight (Gray, 1939; Lawry, 1970). The cycle of activity of a parapodium commences with a slow forward movement of the parapodium, by the end of which movement the tip of the parapodium is applied to the substratum and the setae are protracted to gain a foothold. There follows a more rapid power stroke in which the contraction of parapodial flexor muscles exerts a forward pull on the body, and the parapodium remains in contact with the substratum; the stroke terminates with the parapodium directed backwards and there is a pause of variable duration in this position. Retraction of the setae follows, the parapodium is lifted from the substratum and swung more slowly forwards in a recovery stroke. Such a cycle involves a complex sequence of motor instructions. Neural circuits responsible for controlling these cyclic sequences presumably exist at each side of each segment. The parapodia at opposite sides of a segment perform their power strokes 180° out of phase. The power stroke of one parapodium is followed after an interval by the power stroke of the ipsilateral parapodium of the next anterior segment.

When a polychaete is stimulated to increased activity, the rate of propagation of the waves of parapodial stepping increases markedly, but the frequency of generation of the waves increases little if at all. Faster movement occurs because the body develops lateral undulatory waves, which pass forwards in synchrony with the waves of parapodial activity and the body waves enhance the efficiency of parapodial strokes (Clark & Tritton, 1970). Zones of contraction of longitudinal body muscles alternate at the two sides of the body, the contractions of longitudinal muscles at one side of any segment coinciding in time with the forward recovery stroke of the parapodium on the same side of that segment and with the rearward power stroke of the parapodium and relaxation of longitudinal muscles on the opposite side of the same segment. Reversed swimming is accompanied by a reversal of the direction of propagation of the body undulations.

The parapodia of a walking *Nereis* exhibit a three-dimensional beat cycle; superimposed on the forwards and backwards movement of the limb are vertical ones, in which the parapodial tip is raised and lowered relative to the substratum and to the body. The actively swimming nereid exhibits only a two-dimensional backwards and forwards motion of parapodia. An extra degree of nervous control is therefore required when walking to perform the raising and lowering actions in synchrony with the recovery and power strokes of the limb cycle.

If the body of either *Nereis* (Gray, 1939) or *Harmothoë* (Lawry, 1970) is cut into lengths of ten segments or so, each section of the worm will generate its own waves of parapodial stepping rhythm that pass forwards from the most posterior segment. If the ventral nerve cord is cut with minimal other damage, the parapodial waves are interrupted at the point of cutting, although coordinated waves occur at both sides of the cut. Cutting off one or several parapodia distal to the parapodial ganglion does not interfere with the propagation of either parapodial waves or body undulations. A rhythm of bursts of nervous activity coinciding with the stepping rhythm of parapodia could be detected in the segmental nerve between the nerve cord and the parapodial ganglion of *Harmothoë*, and this nervous rhythm persisted if the nerve was cut between the recording electrode and the parapodial ganglion. Cutting this nerve between the nerve cord and the parapodial ganglion prevents stepping activity of the parapodium concerned although some local reflexes persist in the parapodium (Wilson, 1960). If the nerve between the nerve cord and the parapodial ganglion is cut at one side in several adjacent segments parapodial waves from the posterior on that side of the body are blocked and new waves of ipsilateral parapodial stepping are generated and pass forward from the damaged segments; the waves at the contralateral side pass the region without interruption. Cutting those parapodial nerves does not interfere with the propagation of body undulations, but body waves are blocked by cutting the segmental nerve (IV) which innervates the longitudinal muscles.

These observations suggest that an oscillatory system exists at each side in the nerve cord of each segment which is capable of driving both a system of neural circuits, probably in the parapodial ganglion, which can generate the complete pattern of motor impulses necessary for the performance of a parapodial stroke cycle, and a second system of neural circuits that can control muscle contractions for a reverse stroke cycle. It is not clear whether the same oscillators also provide timing information for the contraction of longitudinal body muscles of each segment; on balance, because cutting segmental parapodial nerves is observed to interfere with parapodial waves without affecting body waves, and because parapodial waves can occur without body waves, it seems likely that separate oscillators are involved, but are closely coupled either by direct nerve connections or by proprioceptive links. The operation of the two parapodial oscillators of a segment 180° out of phase suggests some inhibitory interaction to maintain the phase difference; the passage of waves at one side when none were passing at the other side would support inhibitory rather than excitatory interaction. The requirement for intact nerves between the nerve cord and parapodial ganglion for forward propagation of waves to the next segment suggests that each oscillator normally receives timing information by nerve impulses from the next posterior ipsilateral parapodial ganglion, although observations on isolated sections of the body show that an oscillator can generate its own rhythm in the absence of such timing input. In normal forward movement the activity of a segment regulates the frequency of stepping of the segment in front of it, so that the most posterior segment acts as a pacemaker. A rate command system is present, which exerts an action throughout the body to enhance the propagation of waves along the body, apparently by reducing the time delay between the cycle in one segment and that in the next, reducing the phase difference between the oscillators without having a marked effect on their frequency. Finally, there is also a direction control system which can influence whether forward or reverse parapodial motor circuits are excited by the segmental oscillator and determine whether intersegmental timing pulses are passed to the segment in front or to the segment behind. A diagram suggesting the relationship of adjacent parapodial oscillators, intersegmental coordinating neurons, rate (phase regulation) command and direction command neurones is given in Fig. 6.

The need for metachronism

Among the various locomotor systems discussed here, most walking and all swimming systems show metachronism of the component propulsive structures, the major exception among walking appendages being the tube feet of echinoderms. Metachronism might be expected therefore to be advantageous for the efficient production of locomotion in these multicomponent systems. Two reasons why metachronism might be advantageous are: 1. it enhances the propulsive efficiency of the system, 2. it reduces the likelihood of detrimental interference between adjacent structures.

Unlike the relatively rigid arthropod appendages, cilia and tube feet are

Fig. 6. A schematic hypothesis of the control and coordination of the parapodia of a polychaete worm. The rate command system regulates the frequency of the oscillator circuit, which provides timing signals for the motor circuits (M) that control the limb musculature. A direction command system determines the pattern of motor output from the motor circuits so as to influence the timing of contractions responsible for power strokes and recovery strokes, and also influences whether the forward-conducting or backward-conducting coordination neurons are active in providing phase regulating signals to adjacent segments. The cyclical activity of motor neurons controlling longitudinal muscles would continue in parallel with this system.

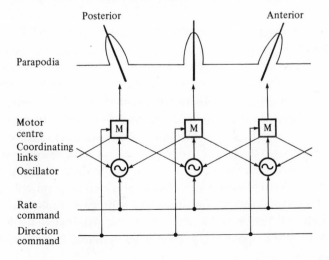

flexible and are able to move between adjacent structures, thereby avoiding serious interference. In most ciliary systems, however, the structures are required to propel water and have highly developed metachronal activity. At any instant in such a metachronal system there are always some structures providing an accelerating force on the water while others are recovering; the result is a fairly continuous flow, the maintenance of which could provide the necessary advantage for metachrony to be adopted by ciliary systems which propel water (Sleigh & Aiello, 1972).

In some other ciliary systems, like the ventral surfaces of planarians, the frontal surface of *Mytilus* gill or vertebrate ciliated epithelia, the cilia are short and very densely packed. Such cilia might be expected to show very high degrees of coordination to avoid interference, whereas in fact metachronism in these cases is poorly developed. The reason for this is probably that these cilia are required to move mucus rather than water, and that with this 'more solid' substratum the advantage of metachrony is reduced.

Echinoderm tube feet also act on a solid surface and are flexible enough to move past each other. For these reasons metachronal coordination does not appear to be necessary for these animals. Indeed there may be a real advantage in not having coordination, for, when moving over uneven surfaces, the independent movement of limbs might only be possible with irregular stepping.

Swimming crustaceans need to propel water and may use multi-limb systems which exhibit metachronism. The metachronal pattern used is similar to that of such ciliated systems as ctenophores, and the maintenance of continuous flow over the limb system is again achieved. Our unpublished observations on tethered *Artemia* illustrate the increased efficiency of propulsion achieved when the water is progressively accelerated by metachronally coordinated structures. As a member of a metachronal system a limb is able to accelerate water to a final velocity greater than that which can be achieved when it is isolated. An equivalent water propulsion to that attainable by a single limb could be achieved by a metachronal system beating at a much lower frequency than the single limb. Also, there is evidence that the force exerted on individual limbs is far less if they are incorporated in metachronal systems, so that less rigid and therefore lighter limbs may be adequate. The rather delicate leaf-like limbs of anostracans are constructed so as to be fairly rigid in the power stroke and flexible in the recovery stroke; this could be related with the absence of reversal in these forms, because reversal would demand modified articulations and increased musculature to maintain stiffness in a reversed effective stroke – the whole limb would need to be more bulky and efficiency would be reduced. Alternative methods of turning to swim in the opposite direction have therefore been developed.

The parapodium of a polychaete worm, such as *Nereis*, does not appear to bend during the swinging movements of the beat cycle, so that reversal of propulsive action of parapodia during reversed swimming does not present the problems found in anostracans. During the recovery phases of the beat cycle the parapodia are bunched closely together at the inside of a curve of the body where they create little drag, whilst at the outside of the curve water is accelerated sequentially by the metachronally coordinated parapodia.

A number of reasons may have led to the development of the multi-segmented, multi-limbed body form found in the Chilopoda and Diplopoda. The use of many legs can be particularly useful for burrowing or pushing through leaf litter. Millipedes with more legs are able to exert more force than those with fewer legs, assuming that the leg length is the same and the segment mass does not alter (Manton, 1953). Longer legs exert bigger forces but would be undesirable in a burrowing animal; the body therefore has tended to take on the form of many short segments, each with short limbs. The small spacing between the legs makes strict coordination necessary if interference between adjacent limbs is to be avoided. Many of the terrestrial arthropods have fewer longer legs which are very strong, but more widely spaced, enabling other forms of gait to be used efficiently in more open habitats. The incorporation of a particular limb into a metachronal sequence does not make that limb more effective in actually pushing against the substratum, because the tip of a limb is less likely to slip relative to a solid substratum, and forward velocity can equal limb-tip velocity. This is not the case in a swimming animal because there is much slippage between the limb tip and the surrounding fluid substratum and only a sequence of successive accelerations given by the component structures of a series can produce a fairly constant velocity that approaches the velocity of the propulsive structures themselves.

Thus, metachronism confers a positive advantage in propulsion of multi-appendage swimming animals and physiological mechanisms have been developed to exploit this advantage. In terrestrial arthropods, however, the multi-limb structure seems to have been developed for physiological or behavioural reasons, so that metachronism may have reduced functional significance, being necessary to provide a means of coordination which prevents interference between limbs. It is interesting that most of the examples of metachronism in the animal kingdom involve propulsion of water.

Parts of this work have been supported by a grant from the Science Research Council. We thank Andrea Morrish for assistance.

References

Aiello, E. (1974). Control of ciliary activity in Metazoa. In *Cilia and Flagella*, pp. 353–76. ed. M. A. Sleigh. London: Academic Press.

Aiello, E. & Sleigh, M. A. (1972). The metachronal wave of lateral cilia of *Mytilus edulis*. *J. Cell Biol.* **54**, 493–506.

Barlow, D. I. & Sleigh, M. A. (1979). Freeze substitution for preservation of ciliated surfaces for Scanning Electron Microscopy. *J. Microsc.* **115**, 81–95.

Cannon, H. G. (1933). On the feeding mechanism of the Branchiopoda. *Phil. Trans. R. Soc., Ser. B*, **222**, 267–352.

Clark, R. B. & Tritton, D. J. (1970). Swimming mechanisms in nereidiform polychaetes. *J. Zool., Lond.* **161**, 257–71.

Cleveland, L. R. & Cleveland, B. T. (1966). The locomotory waves of *Koruga, Deltotrichonympha* and *Mixotricha*. *Arch. Protistenkd.* **109**, 39–63.

Delcomyn, F. (1977). Coordination of invertebrate locomotion. In *Mechanics and Energetics of Animal Locomotion*, ed. R. McN. Alexander & G. Goldspink, pp. 82–114. London: Chapman and Hall.

Eastham, L. E. S. (1934). Metachronal rhythms and gill movements of the nymph of *Caenis horaria* (Ephemeroptera) in relation to water flow. *Proc. R. Soc., Ser. B,* **115**, 30–48.

Eastham, L. E. S. (1936). The rhythmical movements of the gills of nymphal *Leptophlebia marginata* (Ephemeroptera) and the currents produced by them in water. *J. exp. Biol.* **13**, 443–9.

Galt, C. P. & Mackie, G. O. (1971). Electrical correlates of ciliary reversal in *Oikopleura*. *J. exp. Biol.* **55**, 205–12.

Goldstein, S. F. (1974). Isolated, reactivated and laser-irradiated cilia and flagella. In *Cilia and Flagella*, ed. M. A. Sleigh, pp. 111–30. London: Academic Press.

Gray, J. (1928). *Ciliary Movement*. Cambridge University Press.

Gray, J. (1939). Studies in animal locomotion. VIII. The kinematics of locomotion of *Nereis diversicolor*. *J. exp. Biol.* **16**, 9–15.

Gray, J. (1968). *Animal Locomotion*. London: Weidenfeld and Nicolson.

Holwill, M. E. J. & McGregor, J. L. (1976). Effects of calcium on flagellar movement in the trypanosome *Crithidia oncopelti*. *J. exp. Biol.* **65**, 229–42.

Horridge, G. A. (1965). Relations between nerves and cilia in ctenophores. *Am. Zool.* **5**, 357–75.

Hughes, G. M. & Wiersma, C. A. G. (1960). The coordination of swimmeret movements in the crayfish *Procambarus clarkii* (Girard). *J. exp. Biol.* **37**, 657–70.

Hyams, J. S. & Borisy, G. G. (1978). The isolated flagellar apparatus of *Chlamydomonas*. *J. Cell Sci.* **33**, 235–53.

Knight-Jones, E. W. (1954). Relations between metachronism and the direction of ciliary beat in Metazoa. *Q. J. microsc. Sci.* **95**, 503–21.

Knight-Jones, E. W. & Macfadyen, A. (1959). The metachronism of limb and body movements in annelids and arthropods. *Proc. Fifteenth Int. Congr. Zool.* **15**, 969–71.

Lawry, J. V. Jnr. (1970). Mechanisms of locomotion in the polychaete *Harmothoë. Comp. Biochem. Physiol.* **37**, 167–79.

Lowndes, A. G. (1933). The feeding mechanism of *Chirocephalus diaphanus*. Prévost, the fairy shrimp. *Proc. zool. Soc. Lond.* **2**, 1093–118.

Machemer, H. (1972). Properties of polarised ciliary beat in *Paramecium. Acta Protozool.* **11**, 295–300.

Machemer, H. (1974). Ciliary activity and metachronism in Protozoa. In *Cilia and Flagella*, ed. M. A. Sleigh, pp. 199–286. London: Academic Press.

Mackie, G. O., Paul, D. H., Singla, C. M., Sleigh, M. A. & Williams, D. E. (1974). Branchial innervation and ciliary control in the ascidian *Corella. Proc. R. Soc., Ser. B*, **187**, 1–35.

Manton, S. M. (1950). The evolution of arthropodan locomotory mechanisms. Part I. The locomotion of *Peripatus. J. Linn. Soc. (Zool.)* **41**, 529–70.

Manton, S. M. (1953). Locomotory habits and the evolution of the larger arthropodan groups. In *Evolution: Symposia of the Society for Experimental Biology, 7*, pp. 339–76. Cambridge University Press.

Naitoh, Y. & Eckert, R. (1974). The control of ciliary activity in Protozoa. In *Cilia and Flagella*, ed. M. A. Sleigh, pp. 305–52. London: Academic Press.

Naitoh, Y. & Kaneko, H. (1972). Reactivated triton-extracted models of *Paramecium*: modification of ciliary movement by calcium ions. *Science, NY,* **176**, 523–4.

Naitoh, Y. & Kaneko, H. (1973). Control of ciliary activities by adenosine-triphosphate and divalent cations in triton-extracted models of *Paramecium caudatum. J. exp. Biol.* **58**, 657–76.

Parducz, B. (1961). Bewegungsbilder über Didinien. *Ann. Historico-naturales Mus. Natl. Hung., Pars Zool.* **53**, 267–80.

Sleigh, M. A. (1956). Metachronism and frequency of beat in the peristomial cilia of *Stentor. J. exp. Biol.* **33**, 15–28.

Sleigh, M. A. (1960). The form of beat in cilia of *Stentor* and *Opalina. J. exp. Biol.* **37**, 1–10.

Sleigh, M. A. (1968). Metachronal coordination of the comb plates of the ctenophore *Pleurobrachia. J. exp. Biol.* **48**, 111–25.

Sleigh, M. A. (1969). Coordination of the rhythm of beat in some ciliary systems. *Int. Rev. Cytol.* **25**, 31–54.

Sleigh, M. A. (1972). Features of ciliary movement of the ctenophores *Beroë, Pleurobrachia* and *Cestus*. In *Essays in Hydrobiology*, ed. R. B. Clark & R. J. Wootton, pp. 119–36. University of Exeter.

Sleigh, M. A. (1974). Metachronism of cilia in Metazoa. In *Cilia and Flagella*, ed. M. A. Sleigh, pp. 287–304. London: Academic Press.

Sleigh, M. A. & Aiello, E. (1972). The movement of water by cilia. *Acta Protozool.* **11**, 265–77.

Smith, J. E. (1950). Some observations on the nervous mechanisms underlying the behaviour of starfishes. In *Physiological Mechanisms in Animal Behaviour: Symposia of the Society for Experimental Biology,* **4**, pp. 196–220. Cambridge University Press.

Smith, J. E. (1966). The form and functions of the nervous system. In *Physiology of Echinodermata*, ed. R. A. Boolootian, pp. 503–11. New York: Interscience Publishers.

Stein, P. S. G. (1974). Neural control of interappendage phase during locomotion. *Am. Zool.* **14**, 1003–16.

Tamm, S. L. (1973). Mechanisms of ciliary coordination in ctenophores. *J. exp. Biol.* **59**, 231–45.

Tamm, S. L. & Horridge, G. A. (1970). The relation between the orientation of the central fibrils and the direction of beat in cilia of *Opalina. Proc. R. Soc., Ser. B,* **175**, 219–33.

Trueman, E. R. & Jones, H. D. (1977). Crawling and burrowing. In *Mechanics and Energetics of Animal Locomotion*, ed. R. McN. Alexander & G. Goldspink, pp. 204–21. London: Chapman and Hall.

Wilson, D. M. (1960). Nervous control of movement in annelids. *J. exp. Biol.* **37**, 46–56.

H. Y. ELDER

Peristaltic mechanisms

Diverse locomotor mechanisms

Amongst metazoans with few or no hard skeletal parts, which we may loosely refer to as soft-bodied animals, a wide variety of locomotor mechanisms is found. Swimming lies outside the scope of the present discussion (see Trueman, this volume, for some aspects) but it is useful to try to classify the various surface and sub-surface locomotor mechanisms found in soft-bodied animals in order to reach a clearer understanding of the mechanism and origins of peristaltic locomotion (Table 1).

Firstly we may cite those animals which move by means of ciliary mechanisms. This primitive means of movement is largely confined to the smaller

Table 1. *Surface and sub-surface locomotor mechanisms in soft-bodied animals. (See text for further information)*

Locomotor mechanism	Some examples of animals which employ it
Ciliary locomotion	Small platyhelminthes, nemerteans and gastropods
Pedal locomotion	Some anthozoans, platyhelminthes and gastropods
Peristaltic locomotion	Earthworms and synaptid holothurians
Undulatory locomotion	*Ammotrypane* and fast nereid locomotion
Ambulatory locomotion	*Aphrodite*, Onycophora and slow nereid locomotion
Caterpillar locomotion	Lepidopteran larvae and aspidochirote holothuria
Looping locomotion	Leeches and geometrid caterpillars
Penetration and terminal anchorage burrowers	*Peachia, Arenicola, Priapulus*

invertebrates such as the smaller platyhelminthes, nemerteans and some gastropods and since cilia have no part in peristalsis will not be discussed further. A fuller consideration will be found in Clark (1964), and Sleigh (this volume) discusses some aspects of the control of ciliary locomotion. All the other forms of movement which we shall consider depend upon muscular contraction.

The next two mechanisms to be considered, pedal and peristaltic locomotion respectively, have in common that they both involve successions of waves of altered muscular contraction which travel along the body. In pedal locomotion the muscles concerned with locomotion are concentrated in a relatively large flattened 'foot' which offers good surface area contact with the substratum and leaves the rest of the body free for other functions and unencumbered by bulky locomotor musculature. This body plan is well adapted to locomotion over a firm surface and is best exemplified by the gastropod molluscs. The relatively few gastropods which have secondarily become adapted to a burrowing mode of life are the exceptions which prove the rule (for discussion see Trueman, 1975, and Trueman & Brown, 1976).

In contrast, all of the body wall musculature is generally involved in peristaltic locomotion, and consequently the body assumes a more cylindrical shape. In turn this tends to minimise the area in contact with a hard substratum and it is not surprising that a wide variety of secondary gripping devices have been developed amongst animals utilising peristaltic locomotion in order to gain better purchase on the substratum (Table 2). The main advantage of the cylindrical body plan lies in that tensions generated around the body wall circumference result in pressure being distributed in all direc-

Table 2. *Some accessory gripping mechanisms employed in conjunction with peristaltic progression*

Gripping mechanisms	Examples of animals which exhibit it
Mucus	Platyhelminthes, nemerteans, annelids, sipunculids, hemichordates, holothuria, etc.
Adhesive rugae	Anthozoans, e.g. *Aiptasia*
Setae and chaetae	Annelids and dipterous larvae
Suckered appendages	Lepidopteran prolegs, leech posterior sucker, holothurian podia
Cuticular flanges	Dipterous larvae and *Priapulus*
Calcareous spicules	Holothuria
Foot processes	Polychaete parapodia, circum-oral tentacles of synaptid holothuria, lepidopteran forelegs
Mouth or jaws	Leech anterior sucker. On occasion, earthworms, dipterous larvae and *Geonemertes*

tions by the hydrostatic fluid skeleton. As Clark (1964) and Trueman (1975) have pointed out this suits an animal well to a burrowing life, surrounded by substratum.

A clear distinction is usually made between the normally axis-centred waves of contraction which pass along the body in peristalsis and waves which progress alternatively along opposite sides of the body producing the characteristic sinusoidal form of undulatory locomotion. Although such oscillations usually involve right and left body wall muscles as in the polychaete *Ammotrypane*, burrowing fish-like into sand, or *Nereis* in fast locomotion, they could be organised on a dorso-ventral pattern, as in leech swimming. Analysis of this type of locomotion may be found in Clark (1964), Gray (1968) and Clark (1976).

Ambulatory locomotion is another type of progression which is readily distinguishable from peristalsis. Animals which have adopted this type of locomotion have developed leg-like appendages which become the main organs of movement and the evolutionary trend is to brace the body as a firm base from which the legs may operate, as was shown by Mettam (1971) for the polychaete *Aphrodite*. Further discussion will be found in Gray (1968).

Caterpillar and looping locomotion would at first sight appear to differ fairly obviously from peristalsis. However, as will be argued below, it seems probable that some examples have arisen in evolution as the logical outcome for animals which persisted in employing peristaltic locomotion for movement on the surface of the substratum, a usage for which we have seen above peristalsis is not particularly suited. Hughes & Mill (1974) and Trueman (1975) give further discussion of caterpillar and looping locomotion.

Finally, we consider penetration and terminal anchorage burrowers. Clark (1964) first suggested that all soft-bodied burrowing animals progressed through the substratum in a fundamentally similar way. Trueman and his co-workers especially, have greatly increased our knowledge of the mechanism of this form of movement through the substratum in which the animal alternately forms a penetration anchor, against which it thrusts forward into the ground, and a terminal anchor, on which the animal then pulls. This burrowing strategy has arisen independently in at least eight different phyla and probably many more times within some groups. It is the natural end point in the locomotion of soft-bodied animals which try to burrow by peristaltic means and while there are many variations on the theme, including some partial exceptions to the rule of alternate formation of penetration and terminal anchors, they serve only to emphasise what a successful strategy this is for soft-bodied burrowers.

Mechanical considerations

The mechanical problems presented by movement in soft-bodied animals are quite different from those encountered by animals having rigid musculoskeletal systems (Alexander, 1968). In the latter, both the range and direction of movement are often prescribed by the form of hard skeletal articulations (e.g. in many arthropod, vertebrate, echinoderm, etc. joints). Because of this the muscles are often organised into discrete blocks and readily separable into prime movers, antagonists and synergists. In soft-bodied animals, on the other hand, contraction of one set of muscles is most commonly antagonised by another, acting via an enclosed fluid system, the hydrostatic skeleton. Since muscles cannot actively lengthen, such fluid skeletal systems are essential components in the locomotion of soft-bodied animals, and an extensive range of external fluids, body fluids and tissues have been 'pressed' into service (see Clark (1964), Chapman (1975) and Jones (1978a) for details). Since fluids, unlike rigid skeletons, transmit pressure in all directions, contraction of one muscle component in a soft-bodied animal affects all the others and this presents special problems in motor control of posture. It means also that muscles must be organised in sheets rather than blocks, or as bands attached to layers of connective tissue or cuticle, rather than to discrete tendons or apodemes. Most commonly, circular and longitudinal body wall muscles are antagonistic. However oblique, transverse, dorso-ventral and other longitudinal muscles may augment the principal layers amongst platyhelminthes, annelids and arthropod larvae (Clark, 1964). The nematodes possess only longitudinal muscles and while the normal form of locomotion is by sinusoidal waves, peristalsis may occasionally be encountered in small terrestrial nematodes (Lee, 1965). Further Seymour (1973) has illustrated how transportive peristalsis in the nematode gut may be achieved by means of peristaltic waves in the body wall. In the errant polychaete *Nephtys* circular muscles are absent, and the longitudinals are antagonised by dorso-ventral muscles only (see Trevor, 1978, for references). Both undulatory and peristaltic locomotion are observed in *Nephtys*.

In soft-bodied animals tonus must be maintained by all or most of the body wall muscles for all of the time that the animal is active, or the body becomes flaccid. This contrasts with the situation in animals with hard skeletal systems in which some of the muscles may be carried in the relaxed state for most of the time (see for example Wardle & Videler, this volume, on fish muscle). The properties of fluid skeletal systems have been studied by Chapman (1950, 1958, 1975) and Jones (1978a). Hydrostatic fluids, totally enclosed by muscle/connective tissue systems, form very efficient pressure-generating structures, and in recent years the application of electronic measuring techniques has yielded a wealth of information about locomotion in soft-bodied animals, at the hands particularly of Trueman and his co-workers (see Trueman, 1975).

Directions of animal and wave progression

Following the elegant analyses of earthworm (*Lumbricus terrestris*) and ragworm (*Nereis diversicolor*) locomotion (Gray & Lissman, 1938; Gray 1939), it was first pointed out by Gray (1939) that locomotion could be achieved by waves of muscular activity passing along the body either in the same, or in the opposite, direction to that of locomotion. Unfortunately synonyms abound in the terminology of locomotor waves. Thus in relation to waves travelling in the same, and in the opposite direction to locomotion respectively, we find the terms positive and negative (Gray, 1968), synkinetic and antikinetic (Wells, 1949; Mettam, 1969), direct and retrograde (Clark, 1964; Trueman & Ansell, 1969; Seymour, 1971a), symplectic and antiplectic (Sleigh, this volume). Whichever terminology is used, it is most important, as Seymour (1971a) has emphasised, to define the terms in relation to the direction of movement and not with respect to head or tail; in the latter event the terminology becomes confusing or contradictory when the animal moves tail first. In recent years the terms direct and retrograde, first defined by Vlès (1907) in relation to gastropod locomotion, have become most used and will be adopted in this contribution.

As Gray (1968) has emphasised, the direction in which an animal moves in relation to the passage of a wave of muscular activity is dependent upon the phase of the cycle at which the body grips the substratum, i.e. at which the *points d'appui* are formed. The basic mechanism by which locomotion is achieved by retrograde and direct peristaltic waves passing along the body is illustrated in Fig. 1a and b. The direction and speed of advance of equivalent body loci is given by the slopes of the fine dashed lines and the direction and rate of travel of the waves by the coarse dashed lines. Movement relative to the ground of a point on the body occurs only during the passage of a wave, i.e. the advance is cyclical. The other, more distended regions of the body form the *points d'appui* and while any one point is stationary in relation to the ground the leading and trailing edges move according to the travel of the locomotor wave, i.e. the *points d'appui* progress relative to the ground.

Locomotion in terrestrial planarians such as *Bipalium*, *Rhynchodemus* and *Microplana*, and in some of the shorter nemerteans, presents an apparent anomaly, for in these animals the retrograde locomotor wave travels at a rate equal to the rate of progression. The *points d'appui* therefore are stationary relative to the ground and the animals appear to move 'through' a series of discrete steps. The explanation lies in the employment of a second propulsive mechanism; simultaneously with the retrograde muscular waves, ciliary metachronal waves travel directly along the sole. Pantin (1950) first described this type of locomotion in *Rhyncodemus* in which the ciliary

Fig. 1. Diagrams showing how locomotion is achieved by a peristaltic wave travelling (*a*) in the opposite direction to that of the animal (retrograde peristalsis) and (*b*) in the same direction as the animal (direct peristaltic progression). The direction and rate of progress of the waves are given by the coarse dashed lines and that of equivalent points on the body surface by the fine dashed lines. In (*a*) the segments may remain of constant volume. In (*b*) the segments must change volume.

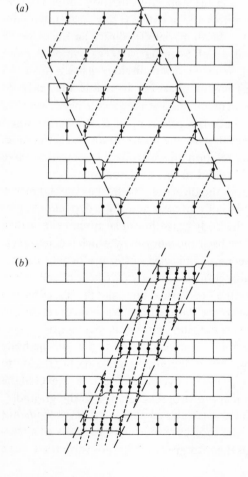

(*a*)

(*b*)

propulsion may be combined with a retrograde pedal wave and further accounts may be found in Clark (1964) and Gray (1968). In the much more extensible species *Microplana britannicus* the retrograde muscular waves are peristaltic (Jones, 1978b) and in terms of the type of diagram in Fig. 1 the coarse dashed lines would become vertical in what Jones (1978b) has called stationary peristalsis.

Presence or absence of segmentation

In most soft-bodied animals employing peristaltic locomotion, the body is primitively non-segmented, as in the Authozoa, Turbellaria, Nemertea and Echinodermata, and in others it has been secondarily lost as in the Mollusca. Amongst these groups the body cavity (or volume of deformable parenchymatous tissue) is continuous through large parts of the body. However, amongst the remaining segmented groups some annelids have a complete system of watertight septa separating the segments, as in the earthworms, while others, such as the lugworm, *Arenicola*, have an 'open' coelom, save for a few specialised septa. The questions have been asked, what are the differences, from the point of view of locomotion, between the segmented and non-segmented states and what are the advantages and disadvantages of each?

From reference to Fig. 1 it may be noted that retrograde peristalsis can occur whether or not the segments are of fixed volume, whereas direct peristalsis can occur only in animals in which there can be fluid displacement from the moving region. Conversely peristaltic waves must be retrograde in an animal with fixed volume segments and in which the action of the longitudinal and circular muscles must be reciprocal. Animals with body segments of variable volume may employ either retrograde or direct peristalsis and the muscle action can be reciprocal or synergic.

Although the merits of the septa in bracing the body wall for generation of radial forces have been discussed by Clark (1964) the difference between a septate and a non-septate state has perhaps best been brought out by Seymour's (1970) comparison, between *Lumbricus* and *Arenicola*, of the radial forces which can be exerted at a 'point' on the body wall. The animals were photographed crawling beneath a narrow perspex bridge device connected to a transducer from which the body wall contact area (A) and force exerted (F) could be measured. Internally measured coelomic pressures (P) times the body wall contact areas (A) were consistently less in both animals than the measured forces (F) (Fig. 2). Seymour (1970) considered these 'excess lifting forces' to be a function of the 'intrinsic rigidity' of the body wall muscle layers. He noted three further points. Firstly, in *Lumbricus*, in which each segment of the body wall is braced by the septa, the higher the coelomic pressure, the more nearly rounded the segmental cross-section becomes and therefore the smaller the area on which the force is exerted. By contrast, in *Arenicola*, where there are no septa in most of the trunk, internal pressure varies directly with body wall area over which it is exerted. Secondly, in the earthworm the body becomes shorter and thicker at a region where maximum lifting force is being generated, thus concentrating more septa in that region, whereas in the lugworm, segments exerting lift were usually simultaneously dilated and elongated, resulting in a locally thin body wall. Thirdly, higher forces are found in *Arenicola* because segments in a large region of the body contract both longitudinal and circular muscles, displacing coelomic fluid which hydraulically expands fewer other segments. In *Lumbricus* the muscles of each segment act reciprocally and pressures are localised by the septa (Seymour, 1976). *Arenicola* is thus well adapted to burrowing in a homogeneous substratum while *Lumbricus* is better adapted to generating the unidirectional lifting forces necessitated by burrowing in a crevice habitat.

Fig. 2. Results of Seymour (1970) after Chapman (1975), showing the relationship between the measured forces (F) in the earthworm *Lumbricus* (E) and the lugworm *Arenicola* (L) and the expected forces calculated from measurements of internal hydrostatic pressure (P) and the area over which they were exerted (A). The difference between these and line a is a measure of the 'intrinsic rigidity' of the body wall.

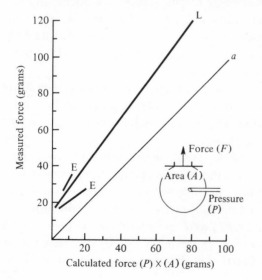

The central nervous system problems of neuronal control and integration are probably very different between an animal with an open body fluid cavity and a completely septate type. In *Arenicola* all of the muscles are involved in every movement in the intercommunicating trunk segments, a feature of animals with open body cavities. (Septa can isolate the posterior tail (Seymour, 1971a) and the head region (Foster-Smith, 1976) from trunk pressure fluctuations.) In *Lumbricus* on the other hand each segment functions essentially independently and electromyographic recordings show that the muscles act reciprocally (Seymour, 1971b). Trueman (1975) comments that comparing pressure records between *Arenicola* and *Lumbricus*, one is left with the impression that the septate state is much more stable. In the trunk region of *Arenicola*, resting pressure fluctuates around 0.5 kPa, while in *Lumbricus* segments it is relatively stable at less than 0.2 kPa, and he speculates that the difference is due to muscle, stretch-receptor feedback originating from the whole trunk musculature in *Arenicola* but from individual segments in *Lumbricus*.

Elder (1972, 1973a) suggested that direct peristaltic progression was an adaptation shown by the polychaete *Polyphysia crassa* to life in very soft flocculent muds. This is confirmed by the finding of similar direct peristaltic waves in animals from similar habitats in at least two other phyla, *Priapulus caudatus* (Elder & Hunter, 1980) and *Leptosynapta roseola* (R. D. Hunter & H. Y. Elder, unpublished). The point may be illustrated with reference to Fig. 1. In the soft mud habitat only relatively low pressures are required and it is of importance to distribute the pressure over *points d'appui* of as large an area as possible. From the scheme in Fig. 1a it can be seen that three segments are involved in the retrograde peristaltic wave, that a point on the body wall surface advances by a distance equal to three dilated body segments during the passage of a wave and that an area of body wall, equivalent to six dilated body segments, is removed from contact with the substratum as the wave passes that region. In scheme 1b the direct peristaltic wave is shown involving four segments which advance by a length, equivalent to two dilated segments and remove a body wall area equivalent to two dilated segments from contact. If the wave involved six segments, in Fig. 1b, a point on the body wall would advance by the same distance as in the retrograde wave of Fig. 1a during the passage of the wave but an area of body wall equivalent to only three dilated segments, i.e. half of that in Fig. 1a, would be removed from contact with the substratum.

Definition and types of peristalsis

It is usually easy to recognise a peristaltic wave passing along an elongated cylindrical animal but there are almost as many variations on the basic theme as there are animals exhibiting peristalsis. A precise description of a particular form of peristaltic wave is difficult to give and has seldom been attempted. Confusion and imprecision are not uncommon in the literature and as a step towards rectifying this, Heffernan & Wainwright (1974) have suggested the following definition. Peristalsis is any muscular contraction moving along a radially flexible tube in such a way that each component wave of circular, longitudinal or oblique muscular contraction is preceded or followed by a period of relative relaxation of all similarly oriented muscle within a given tubular segment. This definition is intended to cover not only locomotor peristalsis but the transportive mechanisms which are common in hollow tubular body organs in many phyla and convey food, urine, blood gametes, etc. The word 'tube' in the definition, which implies a hollow structure, might seem to exclude acoelomate animals such as the nemerteans and turbellarians. Clark (1964) describes a worm as having a flexible, muscular, body wall enclosing an incompressible but deformable medium which has a constant volume and in which fluid pressures can be transmitted. In these terms, animals such as the helminthes, nemerteans and leeches in which much of the 'deformable medium' consists of parenchymatous packing tissues, would also obviously fit the description. The requirement for a more precise description of specific examples of peristaltic mechanisms, as advocated by Heffernan & Wainwright (1974), can best be illustrated by a brief consideration of just some of the diversity of peristaltic locomotor mechanisms which have been described.

In his consideration of the various ways in which peristalsis may be employed to effect locomotion or the irrigation of tubes and burrows, Mettam (1969) recognised the four possibilities shown in Fig. 3. Condition 3a illustrates the fully septate condition with a retrograde wave. The segments are therefore of constant volume and the directions of movement of the animal and the wave are opposite, as we have seen above. This mechanism, retrograde non-overlapping peristalsis in Heffernan & Wainwright's terminology, is exhibited by the earthworms. Fig. 3b illustrates the passage of a direct wave. The direction of wave travel and body advance are the same and the segments must be of variable volume. This type of direct overlapping peristalsis has been described in the polychaete *Polyphysia* (Elder, 1973a) and in synaptid holothuria (Elder, 1973b; Heffernan & Wainwright, 1974). In both of 3a and b, most of the body wall, forming the *points d'appui*, has the circular muscles relatively extended and the travelling waves are seen as localised areas of relative constriction. Heffernan & Wainwright (1974) note that their definition would be equally valid if the word contraction, mentioned twice in their definition, was changed to relaxation and the relaxation was substituted by contraction. The resultant appearance of the peristaltic waves would be as Mettam (1969) illustrated in diagrams 3c and d. Retrograde dilation peristalsis, shown in 3c, has been described for the peristaltic progression exhibited by nemerteans such as *Lineus* and *Amphiporus* (Gray, 1968; Mettam, 1969). The dilated segments as in 3a and b act as the *points d'appui*. If, however, as in the tube-dwelling fan worm, *Sabella*, it is some of the narrow segments which act as the anchorage points, by means of specialised musculature and parapodia, and the shortened segments are lubricated against the tube walls, then the worm stays stationary and the peristaltic dilation forms a piston forcing an irrigatory current through the tube (Mettam, 1969).

The type of peristalsis illustrated in Fig. 3d, direct dilation pristalsis, is found in *Arenicola* (Wells, 1961) and in *Leptosynapta* (R. D. Hunter & H. Y. Elder, unpublished data) when moving in their burrows. These animals, being non-septate, have a variable segmental volume during peristaltic activity and Wells (1961) illustrates both headward, irrigatory pumping in a stationary worm, and tailward creeping, produced by headward waves of

Fig. 3. Diagrams illustrating four types of peristaltic progression: (*a*) retrograde non-overlapping peristalsis; (*b*) direct overlapping peristalsis; (*c*) retrograde dilation peristalsis; (*d*) direct dilation peristalsis. For further explanation see text. Redrawn after Mettam (1969).

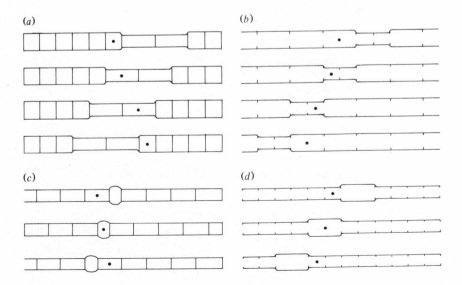

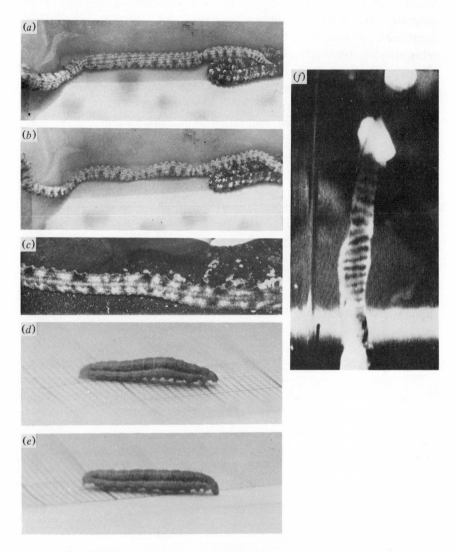

Fig. 4. (*a*) Part of the body of a metre-long synaptid holothurian, *Synapta maculata*, showing the passage of two direct, overlapping peristaltic locomotor waves from right to left. The obviously constricted regions of the body wall are followed by marked dilation and extension (i.e. relaxation of both circular and longitudinal muscles) of the body wall.

    (*b*) The same animal again moving from right to left but now employing a combination of direct overlapping peristalsis and direct arching peristalsis. Between the waves of the former locomotor pattern the body is arched off the substratum by a contraction of the ventral longitudinal muscles.

localised dilation. The former could be achieved by peristalsis involving only a relative relaxation of the circular muscles and no·shortening of the segments. It must involve an increase in segmental volume as the wave passes and is only possible in non-septate animals.

Movements of all four types described in Fig. 3 involve regions of the body which are narrowed and regions which are broadened. It can be seen therefore that our recognition and description of peristalsis relates to the more localised circular muscle state which is then regarded as moving along the body. There is no reason why intermediates between either 3a and 3c or 3b and 3d should not occur and indeed Mettam (1969) points out that the peristalsis of the hemichordate proboscis described by Knight-Jones (1952) is intermediate between types 3a and 3c. A further point to be noted, however, is that in types 3a and 3b the whole body of the animal advances relative to the substratum only when the wave has traversed the entire length of the animal, whereas in 3c and 3d any progression of the wave advances the animal (Mettam, 1969).

Heffernan & Wainwright (1974) clearly regard 'caterpillar locomotion' as a form of peristalsis, which they describe as direct arching peristalsis. The way in which one type of peristalsis may grade into another is illustrated in the three photographs of advancing synaptid holothurians (Fig. 4a–c). Fig. 4a shows the most common form of locomotion found in these metre-long, largely surface-living, synaptids, namely direct overlapping peristalsis. In Fig. 4b, of the same animal, the form of the wave has changed and although circular muscle constriction is still present it is less marked. The abrupt transition of both circular and longitudinal relaxation following the constriction of the direct peristaltic wave (Fig. 4a) is also much less pronounced in Fig. 4b. Additionally a marked arching of the body is seen between the

---

(c) Part of the body of another S. maculata moving from left to right employing only direct arching peristalsis in which no obvious change in the length of the circular muscles is observed.

(d) and (e) show a caterpillar of the small cabbage white butterfly Pieris rapae moving from left to right by typical 'caterpillar' locomotion, direct arching peristalsis. When the segments are detached and arched off the substratum they are 20–30% shorter on the ventral surface than either the following attached segments or the same segments during the attached phase. See text for further details.

(f) An in vitro Magnus preparation of rabbit distal colon in which transverse marks have been dyed onto the serosal surface by the supra-vital stain Janus green B. This photograph is of a frame in a series obtained by a stop-frame videotape recording technique used to study the sequence of length changes of the circular and longitudinal muscles during the passage of a transportive peristaltic wave. See text for further explanation. (W. E. Brodie, H. Y. Elder & V. A. Moss, unpublished).

peristaltic waves in the latter. Transition to a true caterpillar type of wave is complete in the form of locomotor wave seen in Fig. 4c in which circular muscle contraction plays little if any part in the passage of the wave. Instead a wave of dorsal, longitudinal muscle contraction raises the body off the substratum and precedes a wave of ventral longitudinal shortening which largely determines the form of the arch and the distance by which the body wall is advanced by the wave. These large synaptids live mainly on the surface of hard substrates; over firm, sandy ground amongst boulders and eel grass (*Thalassia*) in the case of *Synapta maculata*. Caterpillar locomotion probably offers two advantages to animals moving over a firm surface, namely a potentially greater step size (see below) and absence of frictional drag of the advancing region of body wall. It is not appropriate to burrowing animals and is not found in burrowing synaptids such as *Leptosynapta* (R. D. Hunter & H. Y. Elder, unpublished). Direct overlapping peristalsis has probably arisen from direct arching peristalsis in the holothuria.

The caterpillar locomotion of lepidopterous larvae was studied by Barth (1937) whose functional interpretation of the operation of the complex musculature is marred by an error pointed out in the account of caterpillar locomotion by Hughes & Mill (1974). Barth (1937) rightly illustrates the wave of dorsal longitudinal muscle shortening preceding that of the ventral longitudinal muscles but illustrates the latter as occurring during the first attached segment following the wave. As may be deduced from a theoretical consideration, the shortening of the ventral longitudinal muscles must precede the following *point d'appui*. Measurements from pairs of photographs such as those shown in Fig. 4d and 4e show that the ventral region of segments at the trailing edge of the peristaltic arch (i.e. just prior to reattachment) are 20–30% shorter than those of either the following segment or the same segment measured when no wave was passing that region of the body. Again, using the method of measuring the length of the ventral region of a segment of the body wall during the passage of a direct, arching, peristaltic wave just prior to attachment and comparing it with the length of the same region when forming part of the *point d'appui*, in the aspidochirote holothurian *Bohadschia graffei*, revealed up to 55% shortening during the passage of the wave (H. Y. Elder, unpublished).

Looping locomotion is probably best known and exemplified by the leeches although it has arisen independently in members of at least six phyla. The fundamental similarity of peristalsis and leech-like looping locomotion was acknowledged by Heffernan & Wainwright (1974) who classify both earthworm and leech locomotion as retrograde, non-overlapping peristalsis. Insight into the way in which leech locomotion may have arisen from earthworm-like locomotion is given by the observations of Yapp (1956) on small earthworms ascending the smooth surface of a sheet of glass. The mouth was used as a sucker to increase grip and some of the segments were arched off the surface, to reduce friction. Looping locomotion may equally have derived from pedal locomotion. Trueman (1975) points out that there is merely a degree of difference rather than of kind between retrograde pedal or peristaltic progression and looping locomotion. Reduction of the number of waves passing along the body at any one time reaches the limiting condition of one, and if much of the body is involved in a single wave, which has advantages in terms of step size, there is a tendency to develop accessory gripping mechanisms for the remaining anterior and/or posterior *points d'appui*, e.g. suckers in the leeches and the commensal nemertean *Malacobdella* or adhesive bristles in the draconematid and elipsonematid nematodes (Clark, 1964).

It is misleading however to associate leech locomotion solely with retrograde waves. Certainly the normal resting position of leeches is in the arched (longitudinally contracted phase) and the first act of locomotion is therefore detachment of the anterior sucker and body extension by means of a wave of circular muscle contraction spreading backwards. Amongst the arthropods Hughes & Mill (1974) regarded the locomotion of the looper caterpillars of the geometrid moths as distinct from caterpillar locomotion since it did not involve retrograde locomotion. The main difference however is solely that these larvae normally come to rest in the extended phase of the cycle resembling a twig, attached at the posterior end. The first act of locomotion is to attach with the forelegs, detach the pro-legs and initiate ventral longitudinal muscle contraction from the rear. This exaggerated direct arching peristalsis seems logically to have arisen from more usual caterpillar locomotion.

The basic burrowing mechanism

Clark (1964) proposed that burrowing soft-bodied animals have a common strategy which depends upon alternate formation of a more posterior fixed point or anchor against which the animal thrusts into the substratum and a distal anchor towards which the animal pulls the rest of the body. Based upon data from many more studies, Trueman (1968) extended the concept of the universality of the basic burrowing mechanism based upon formation of alternate 'penetration' and 'terminal' anchors, and has emphasised that, as in the evolution of looping locomotion on the surface, so the basic burrowing mechanism is the logical end point of reduction in the number of pedal or peristaltic waves (Trueman, 1975). An observational aid to the direct study of burrowing mechanisms in animals, which rather surprisingly has been little pursued, was the introduction of the artificial translucent substrate 'Cryolite' (Josephson & Flessa, 1972). Despite the difficulties of observing the mechanisms, and measuring the mechanical performance, of soft-bodied animals burrowing under the surface, the use of electronic measuring devices, particularly by Trueman and his colleagues (see Trueman & Ansell, 1969; Trueman, 1975) has added greatly to our understanding of burrowing. In several species it has been possible to estimate the metabolic efficiency of burrowing and Trevor (1978) has recently collated the available evidence. Using Alexander's (1975) equation for the 'net cost of transport'

$$\text{net cost of transport } (\text{J kg}^{-1}\text{m}^{-1}) = \frac{\text{power during locomotion} - \text{power at rest}}{\text{body mass} \times \text{velocity}}$$

Fig. 5. Plots of the 'net cost of transport' (see text for calculation) against fresh body weight. Data on running (+), flying (×) and swimming (*) animals are by respirometry from Alexander (1975). Data for burrowing animals (○) are from mechanical measurements by Trevor (1978). Solid circles (●) assume an efficiency of chemical conversion of 20% and are probably the more valid comparison with Alexander's data. Redrawn from Trevor (1978).

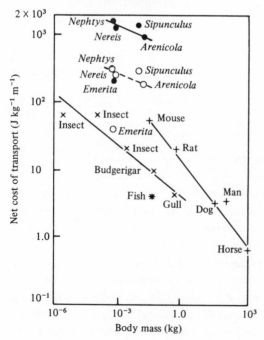

Trevor (1978) plotted the energetic cost of burrowing (Fig. 5) for a range of invertebrates using both a fresh body weight basis (open symbols) and a 20% chemical conversion efficiency factor (solid symbols, Fig. 5). The latter give the more valid comparison with the other data from other forms of terrestrial locomotion obtained by respirometry (Alexander, 1975). From either method of calculation the metabolic cost of burrowing is clearly high and the evolutionary advantages of adopting this habit are certainly other than locomotor. For a further discussion of the locomotor energetics see Trevor (1978) and the chapters by Goldspink and Alexander (this volume).

In his influential and thought-provoking book, Clark (1964) unfortunately concluded that direct peristaltic waves were unsuitable for burrowing invertebrates since the synergic contraction of the localised region in the wave would be opposed by the contraction of all of the rest of the body wall musculature. Direct peristalsis must therefore, of necessity, operate at low hydrostatic skeletal pressures. Examples of animals employing direct overlapping peristaltic waves in burrowing have now been described from four phyla, the anemone *Peachia hastata* (Ansell & Trueman, 1968), the polychaete *Polyphysia crassa* (Elder, 1972, 1973a), the holothurian *Leptosynapta tenius* (Elder, 1973b) and the aschelminth *Priapulus caudatus* (Friedrich & Langeloh, 1936; Hammond, 1970; Elder & Hunter, 1980). In all of these it is confirmed that very low body fluid pressures are employed during the passage of the direct peristaltic wave. In each case it is true that alternative mechanisms are used to dislodge and excavate the substratum and the direct peristaltic wave serves merely to advance the body. In *Peachia* complex movements of the physa disturb and displace the sand. In *Polyphysia* and *Leptosynapta*, lateral scraping of prostomial horns and circum-oral tentacles respectively, achieve lateral displacement and in *Priapulus* the praesoma is forcefully everted to penetrate the substratum. In *Priapulus* the function of the direct peristaltic wave has been an enigma for the last forty years but recent studies (Elder & Hunter, 1980) have emphasised the importance of a mechanism to advance the body wall relative to the animal and to the substratum, without causing the animal to advance relative to the substratum. In *Priapulus*, following the establishment of a terminal anchor the animal pulls forward upon it, leaving much of the body wall concertina'd in narrowed and shortened posterior trunk segments (Fig. 6a). The direct peristaltic wave functions to shift this 'slack' body wall anteriorily and in particular enables the animal to extend into the cavity left by the introverting praesoma (Fig. 6b, c). Without such a mechanism all of the advance into the substratum gained by the penetrating praesoma at eversion would be surrendered again at retraction. Most probably the elaborate burrowing cycle involving the direct peristaltic wave in

Priapulus is an adaptation to living in the soft mud habitat as has been claimed for *Polyphysia* (Elder, 1973a) and for *Leptosynapta roseola* (R. D. Hunter & H. Y. Elder, unpublished data). However it would be surprising if a phase, so necessary in the locomotor cycle of these animals, was totally lacking in animals burrowing in firmer substrata. It is therefore suggested that, at least in some cases, such movements will be found in subsequent studies and have hitherto gone unnoticed or unrecognised. Thus in the much studied *Arenicola*, occasional mention is found of a transient, direct peristaltic wave as a component of some of the digging cycles (Wells, 1949; Seymour, 1971a). Probably this wave serves the very important function of advancing the body wall, but not the animal, relative to the substratum. Nor need this function be served by a direct peristaltic wave. Unpublished observations by Hunter & Elder noted a retrograde wave which served this function in the burrowing sipunculid *Phascolopsis gouldii*.

Despite individual specialisations and elaborations of the burrowing mechanism, which may be imposed by anatomy and mode of life, the validity of the basic burrowing mechanism is established and exceptions only emphasise the rule. Thus in the polychaete *Polyphysia* and the holothurian

Fig. 6. Diagram showing the function of the direct overlapping peristaltic wave in *Priapulus* locomotion. It serves to advance the body wall of the animal with respect to both the animal and the substratum but the animal as a whole is not advanced relative to the ground by the passage of the wave. See text for further details. From Elder & Hunter (1980).

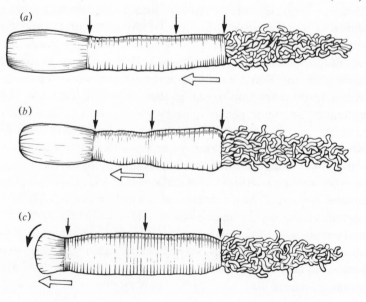

Leptosynapta which possess lateral scraping devices, as mentioned above, no terminal anchor is formed. In these animals the head advances steadily into the substratum and is kept supplied with body wall slack by direct overlapping peristaltic waves (Elder, 1973a). The tail end advances stepwise and such animals always employ a penetration anchor, which therefore appears to be of more fundamental importance than the terminal anchor. That there are so few exceptions to the more normal cyclical advance of the animal into the substratum by alternate penetration and terminal anchorage formation serves to underline the efficiency of the basic mechanism which has evolved independently in many phyla.

Peristalsis and connective tissue organisation

A further consequence of the utilisation of direct peristaltic waves involving synergic action of circular and longitudinal muscle layers, as distinct from the reciprocating action more usual with retrograde waves, is that if the body wall tissues are of constant volume then there must be a mechanism to accommodate and control the large radial distension which will occur during the passage of the wave. An elaborate dermal architecture for achieving this has been demonstrated in *Polyphysia* (Elder, 1972) and in several other invertebrates including the Onycophora (Owen, 1959) and the synaptid holothuria (Elder, 1973b). An alternative solution to the problem may exist in the body wall structures of *Priapulus* and the polychaete *Scalibregma inflatum* in which synergic muscle contraction may result in fluid displacement from between muscle blocks, rather than radial extension.

Recently some of these methods and ideas have been extended to the study of transportive peristalsis in mammalian gut. Changes in longitudinal and circular muscle length can be followed during the passage of peristaltic waves in lengths of gut on which 'segments' have been marked by means of supra-vital dyes such as Janus green B (Fig. 4*f*). Movements in time sequence, are recorded on videotape with closed circuit television and individual frames at the desired time intervals are taken, by stop-frame facilities, for analysis by computer, image-analysis methods. It is now possible to resolve the long-standing debate as to whether the intestinal muscle action is synergic (Bayliss & Starling, 1899; Bortoff & Ghalib, 1972) or sequential (Raiford & Mulinos, 1934; Gonella, 1972) by demonstrating simultaneous activity, at least in rabbit distal colon (Brodie, Elder & Moss, unpublished). Since it is clearly possible to produce transportive peristalsis by means of circular muscles alone it is further suggested that a function of the accompanying longitudinal shortening is to increase the radial thickening of the gut wall locally, as an additional aid to occlusion of the lumen and an aid to propulsion of the bolus.

In conclusion it may be stated confidently that much of interest remains to be elucidated both in the field of invertebrate peristaltic locomotion and the extension of observations and experiments to other forms of peristalsis such as transportive peristalsis.

Original work reported here was supported by the SRC and some observations were made during the α-Helix South-East Asia Bioluminescence Expedition 1975, supported by the NSF.

References

Alexander, R. McN. (1968). *Animal Mechanics*. London: Sidgwick & Jackson.

Alexander, R. McN. (1975). *Biomechanics*. Outline Studies in Biology Series. London: Chapman & Hall.

Ansell, A. D. & Trueman, E. R. (1968). The mechanism of burrowing in the anemone, *Peachia hastata* Gosse. *J. exp. mar. Biol. Ecol.* **2**, 124–34.

Barth, R. (1937). Muskulatur und Bewegungsart der Raupen zugleich ein Beitrag zur Spannbewegung und Schruckstellung der Spannerraupen. *Zool. Jahrb. (Anat.).* **62**, 507–66.

Bayliss, W. M. & Starling, E. H. (1899). The movements and innervation of the small intestine. *J. Physiol., Lond.* **24**, 99–143.

Bortoff, A. & Ghalib, E. (1972). Temporal relationship between electrical and mechanical activity of longitudinal and circular muscle during intestinal peristalsis. *Am. J. dig. Dis.* **17**, 317–25.

Chapman, G. (1950). Of the movement of worms. *J. exp. Biol.* **27**, 29–39.

Chapman, G. (1958). The hydrostatic skeleton in the invertebrates. *Biol. Rev.* **33**, 338–71.

Chapman, G. (1975). Versatility of hydraulic systems. *J. exp. Zool.* **194**, 249–70.

Clark, R. B. (1964). *Dynamics in Metazoan Evolution.* Oxford: Clarendon Press.

Clark, R. B. (1964). *Dynamics in Metazoan Evolution.* Oxford: *Perspectives in Experimental Biology*, vol. 1, ed. P. Spencer Davies, pp. 437–46. Oxford: Pergamon Press.

Elder, H. Y. (1972). Connective tissues, body wall structure, and their significance for the polychaete *Polyphysia crassa* (Lipobranchius Jeffreysii) (Oersted). *J. mar. biol. Ass. UK*, **52**, 747–64.

Elder, H. Y. (1973a). Direct peristaltic progression and the functional significance of the dermal connective tissues during burrowing in the polychaete *Polyphysia crassa. J. exp. Biol.* **58**, 637–55.

Elder, H. Y. (1973b). Distribution and functions of elastic fibers in the invertebrates. *Biol. Bull. mar. biol. Lab., Woods Hole*, **144**, 43–63.

Elder, H. Y. & Hunter, R. D. (1980). Burrowing of *Priapulus caudatus* and the significance of the direct peristaltic wave. *J. Zool., Lond.* **191** (in press).

Foster-Smith, R. L. (1976). The rôle of the head coelon in proboscis eversion in Arenicola marina (L). *J. exp. mar. Biol. Ecol.* **23**, 85–96.

Friedrich, H. & Langeloh, H-P. (1936). Untersuchungen zur Physiologie der Bewegung und des Hautmuskelschlauches bei *Halicryptus spinulosus* und *Priapulus caudatus*. *Biol. Zent.* **56**, 249–60.

Gonella, J. (1972). Modifications of electrical activity of the rabbit duodenum longitudinal muscle after contractions of the circular muscle. *Am. J. dig. Dis.* **17**, 327–32.

Gray, J. (1939). Studies in animal locomotion VIII. The kinetics of locomotion of *Nereis diversicolor*. *J. exp. Biol.* **16**, 9–17.

Gray, J. (1968). *Animal Locomotion*. London: Weidenfeld & Nicolson.

Gray, J. & Lissmann, H. W. (1938). Studies in animal locomotion. VII Locomotory reflexes in the earthworm. *J. exp. Biol.* **15**, 506–17.

Hammond, R. A. (1970). The burrowing of *Priapulus caudatus*. *J. Zool., Lond.* **162**, 469–80.

Heffernan, J. M. & Wainwright, S. A. (1974). Locomotion of the holothurian *Euapta lappa* and redefinition of peristalsis. *Biol. Bull. mar. biol. Lab., Woods Hole*, **147**, 95–104.

Hughes, G. M. & Mill, J. P. (1974). Dynamics of locomotion: terrestrial. In *The Physiology of Insecta*, 2nd edn, vol. III, ed M. Rockstein, pp. 372–9. New York & London: Academic Press.

Jones, H. D. (1978a). Fluid skeletons in aquatic and terrestrial animals. In *Comparative Physiology*, ed. K. Schmidt-Nielsen, pp. 267–81. Cambridge University Press.

Jones, H. D. (1978b). Observations on the locomotion of two British terrestrial planarians (Platyhelminthes, Tricladida). *J. Zool., Lond.* **186**, 407–16.

Josephson, R. K. & Flessa, K. W. (1972). Cryolite: A medium for the study of burrowing aquatic organisms. *Limnol. Oceanogr.* **17**, 134–5.

Knight-Jones, E. W. (1952). On the nervous system of *Saccoglossus cambriensis* (Enteropneusta). *Phil. Trans. R. Soc., Ser. B*, **236**, 315–54.

Lee, D. L. (1965). *The Physiology of Nematodes*. Edinburgh: Oliver & Boyd.

Mettam, C. (1969). Peristaltic waves of tubicolous worms and the problem of irrigation in *Sabella pavonina*. *J. Zool., Lond.* **158**, 341–56.

Mettam, C. (1971). Functional design and evolution of the polychaete *Aphrodite aculeata*. *J. Zool., Lond.* **163**, 489–514.

Owen, G. (1959). A new method for staining connective tissue fibres, with a note on Liang's method for nerve-fibres. *Q. J. Microsc. Sci.* **100**, 421–4.

Pantin, C. F. A. (1950). Locomotion in British terrestrial nemertines and planarians: with a discussion on the identity of *Rhyncodemus bilineatus* (Mecznikow) in Britain and on the name *Fasciola terrestris* O. F. Muller. *Proc. Linn. Soc. Lond.* **162**, 23–37.

Raiford, T. & Mulinos, M. G. (1934). The myenteric reflex as exhibited by the exteriorized colon of the dog. *Am. J. Physiol.* **110**, 129–36.

Seymour, M. K. (1970). Skeletons of *Lumbricus terrestris* L. and *Arenicola marina* (L). *Nature, Lond.* **228**, 383–5.

Seymour, M. K. (1971a). Burrowing behaviour in the European lugworm *Arenicola marina* (Polychaeta: Arenicolidae). *J. Zool., Lond.* **164**, 93–132.

Seymour, M. K. (1971b). Coelomic pressure and electromyogram in earthworm locomotion. *Comp. Biochem. Physiol.* **40A**, 859–64.

Seymour, M. K. (1973). Motion and the skeleton in small nematodes. *Nematologica*, **19**, 43–8.

Seymour, M. K. (1976). Pressure difference on adjacent segments and movement of septa in earthworm locomotion. *J. exp. Biol.* **64**, 743–50.

Trevor, J. H. (1978). The dynamics and mechanical energy expenditure of the polychaetes *Nephtys cirrosa*, *Nereis diversicolor* and *Arenicola marina* during burrowing. *Estuarine Coastal mar. Sci.* **6**, 605–19.

Trueman, E. R. (1968). Burrowing habit and the early evolution of body cavities. *Nature, Lond.* **218**, 96–8.

Trueman, E. R. (1975). *The Locomotion of Soft-bodied Animals*. Contemporary Biology Series. London: Edward Arnold.

Trueman, E. R. & Ansell, A. D. (1969). The mechanism of burrowing into soft substrata by marine animals. *Oceanogr. mar. Biol. Annu. Rev.* **7**, 315–66.

Trueman, E. R. & Brown, A. C. (1976). Locomotion, pedal retraction and extension, and the hydraulic systems of *Bullia* (Gastropoda: Nassaridae). *J. Zool., Lond.* **178**, 365–84.

Vlès, F. (1907). Sur les ondes pédieuses des Mollusques raptateurs. *C. r. hebd. Séances Acad. Sci., Paris,* **145**, 267–8.

Wells, G. P. (1949). Respiratory movements of *Arenicola marina* L: intermittent irrigation of the tube, and intermittent aerial respiration. *J. mar. biol. Assoc. UK,* **28**, 447–64.

Wells, G. P. (1961). How lugworms move. In *The Cell and the Organism*, ed. J. D. Ramsey & V. B. Wigglesworth, pp. 209–33. Cambridge University Press.

Yapp, W. B. (1956). Locomotion of worms. *Nature, Lond.* **177**, 614.

E. R. TRUEMAN

Swimming by jet propulsion

Introduction

Jet propulsion is utilised by a variety of aquatic animals both for normal swimming, by jellyfish and squid, and for escape movements by some dragonfly larvae (Hughes & Mill, 1966), the opisthobranch *Notarchus* (Martin, 1966) and some squid. Apart from dragonfly larvae and some fish

Fig. 1. Generalised diagram summarising events of a single jet cycle of approximately 1 s duration. Muscle, represents contraction during the power stroke followed by restoration during recovery either by antagonistic muscle as in cephalopods or by elastic storage energy as in medusae, scallops or salps. Force curves after Johnson *et al.* (1972).

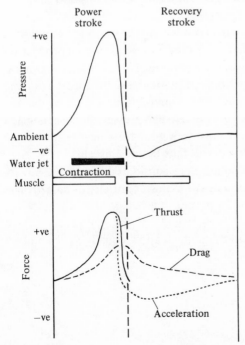

(Lagler, Bardach & Miller, 1962), this means of swimming is characteristically a development of soft-bodied invertebrates. A common principle applies to all animals using jet propulsion: the expulsion of water in one direction to propel the animal in the opposite direction. To accomplish this a body form is required which allows a considerable quantity of water to be expelled (power stroke, Fig. 1) and which is then able to expand, inhaling water for a further power stroke. The effectiveness of a mechanism for jet propulsion is largely dependent on: (i) the velocity and mass of water ejected; (ii) the mass of the animal; and (iii) the magnitude of the drag forces derived from rapid motion in water. The first of these is related to the strength of the musculature expelling water, the cross-sectional area of the jet aperture and the capacity of the chamber from which water is expelled. The relatively great capacity of the mantle cavity and strong muscles of the squid *Loligo vulgaris*, a powerful swimmer, have been contrasted with the relatively smaller cavity and weaker muscles of *Octopus* (Trueman & Packard, 1968). For unit body weight the greatest thrust is developed by the largest chamber with the most powerful musculature and in any animal in which these are constant the duration of the jet pulse is dependent on the jet aperture. Maximum aperture and pressure may produce the greatest thrust but at the high velocities then attained the animal rapidly becomes retarded by drag forces. During normal swimming the aperture is constricted to produce a pulse of longer duration, slower motion and less drag.

Jet propulsion of animals is a cyclical event, with the possible exception of *Pyrosoma* (Weihs, 1977), there are no instances of the continual production of a water jet. Weihs (1977) discusses the advantages of the periodic generation of a jet and concludes that periodic pulses increase the thrust available from a jet of given mass. Each jet cycle is divisible into the two distinct phases of power and recovery strokes (Fig. 1), high pressure being generated during the former to expel water, whilst during recovery inspiration occurs and the pressure in the fluid-filled cavity falls below ambient.

The power stroke and inhalation are here considered before the advantages and disadvantages of this form of swimming are briefly discussed. The reader is referred to articles on the swimming of fish, e.g. Alexander (1978); Wardle & Videler (this volume) for accounts of the hydrodynamics of motion in a fluid, the main principles of which apply equally to fish and jet propelled animals.

The power stroke

Cephalopoda

Jet propulsion is best developed in the Cephalopoda and it is upon these animals that most research has been carried out (Trueman, 1975). In coleoid cephalopods the approximately cylindrical mantle cavity lies postero-ventrally to the head and viscera and its thick muscular wall consists largely of circular muscle fibres which contract to effect the power stroke (Fig. 2). Reduction of the circumference of the mantle causes a decrease in the volume contained within the mantle cavity proportional to the linear change squared. A volume directly proportional to the shortening of the cavity would be ejected as a result of reduction in length (Ward & Wainwright, 1972). The circular muscles are thus well suited to provide the jet forces and in *Loligo* they contract by about 30% for a pressure pulse of 30 kPa at a tensile stress of about $1.5 \times 10^5 \, \text{N m}^{-2}$ (Trueman & Packard, 1968). The inhalant and exhalant openings are separate and a unidirectional flow is achieved by a valve in the funnel and the dilation of the membrane at the base of the funnel to occlude the inhalant apertures during the power

Fig. 2. Diagram of squid in longitudinal and transverse sections showing mantle cavity (m.c.) viscera (v.) and detail of the structure of the mantle musculature. After Packard & Trueman (1974).

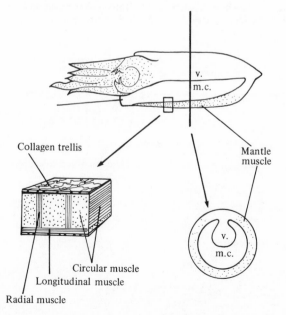

stroke. The motion of a squid has been discussed by Johnson, Soden & Trueman (1972) with particular reference to inertial forces and jet thrust and it is sufficient here to refer to an expression for jet thrust ($q\,\mathrm{d}m/\mathrm{d}t$, where q is jet velocity, m, exhaled water and t, time) $q(\mathrm{d}m/\mathrm{d}t) = 2\,C_\mathrm{D}aP$ where C_D is a coefficient of discharge , a, the area of the funnel aperture and P, the pressure in the mantle cavity. Thus jet velocity and thrust may be determined provided funnel aperture, amount of water expelled, pressure in the mantle cavity and duration of the jet pulse can be measured. The latter are readily accomplished by attaching the squid by a thread to a force transducer which allows recordings to be made of jet thrust and pressure simultaneously while the animal is almost motionless (Fig. 3). This technique has the advantage that the effect of drag and of carried surface water (Prandlt & Tietgens, 1957) need not be considered in making initial comparisons of jet performance in different animals. Jet thrust and pressure are compared in Fig. 3 and the difference between the regression line and the theoretical values is some measure of the coefficient of discharge. However these values are dependent on the accurate measurement of the funnel aperture (a)

Fig. 3. Graph showing relation between pressure and jet thrust in *Sepia officinalis* and *Octopus vulgaris* with diagram indicating manner of recording these simultaneously by Statham transducer and cannula to mantle cavity (P) and by force transducer (F) with thread attached to anterior of animal. Broken line in each example indicates relationship thrust = $2C_\mathrm{D}\,aP$ where C_D is 1, continuous line represents regression line for each species. After Trueman & Packard (1968).

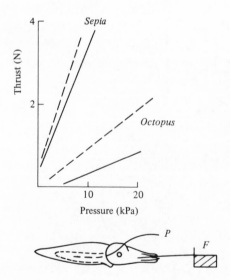

during jetting. This is difficult to achieve and the low value for *Octopus* may be due to constriction of the funnel as the water is expelled.

Both *Loligo* and *Sepia* are capable of powerful swimming but pre-eminent amongst animals swimming by jet propulsion are oceanic squid of the genus *Ommastrephes*, which commonly appear to jet with pressures in excess of 50 kPa (M. R. Clarke and E. R. Trueman, unpublished).

Recent observations, using a catheter tip pressure transducer inserted in the mantle cavity of *Sepia* and the squid *Alloteuthis subulata*, have shown that the form of the pressure pulse is not always simple as had been previously thought (Fig. 4*b*, *c*), *Sepia*, for example, produces a spike of variable amplitude superimposed upon a lower amplitude pulse (M. R. Clarke and E. R. Trueman, unpublished). These additional spikes generally only occur when the animal is particularly fresh and produce a volley of pulses at frequencies of up to 3 Hz. Their cause could be either a second contraction of the mantle muscles or the contraction of the retractor muscles of funnel or head. The absence of evidence of any second contraction of mantle muscle and slight movement of the head during jet pulses, as observed on cine film of jetting, suggests the latter possibility is likely to be true. It may be noted that the head retractor muscles are served by second-order giant nerve fibres from the brain as are the stellate ganglia (Young, 1939).

Many coleoid cephalopods are well adapted for jet swimming because of their streamlined shape and nearly cylindrical mantle cavity. The loss of an external shell during evolution has allowed this development. The only living cephalopod with a true external shell is *Nautilus*, which must preclude fast movements by jetting, nevertheless recent investigations indicate pressure pulses of the order of 2.6 kPa in the funnel (A. Packard, Edinburgh University, personal communication). Such pulses are considered to be generated by the head retractor muscles rather than by mantle muscles which must in any case be limited by the role of the mantle in lining the body chamber of the shell.

Medusae and siphonophore bells

Medusae swim by rhythmic contraction of the bell, so expelling water downwards. Each jet cycle consists of the alternating antagonistic contraction of the subumbrellar muscles (power stroke) and the reaction of the elastic mesoglea. During recovery water is drawn into the subumbrellar cavity from below for jellyfish have a common jet and inhalant aperture. Gladfelter (1972a, b) has compared swimming in a cup-shaped hydromedusan, *Polyorchis*, and a disc-shaped scyphomedusan, *Cyanea*. He showed, by analysis of cine films of *Polyorchis*, that there was a small residual velocity at

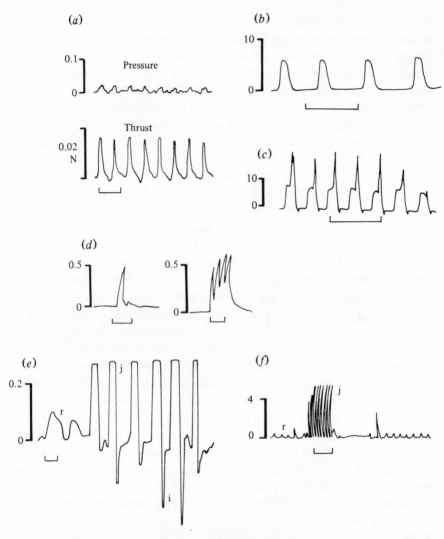

Fig. 4. Extracts of traces showing pressure pulses (kPa) of jetting (j) and respiration (r). Time scale 1 s except in (*f*) where it represents 5 s. (*a*) *Carybdea alata*, simultaneous recording of pressure from subumbrellar cavity and thrust. (*b*) *Sepia officinalis*, succession of pressure pulses from mantle cavity after initial rapid volley (*c*). (*c*) *S. officinalis* jet pulses with superimposed spikes recorded during initial rapid volley (*b* and *c*, M. R. Clarke & E. R. Trueman, unpublished). (*d*) *Chlamys opercularis*, traces of single pulse and volley, from mantle cavity (from Moore & Trueman (1971). (*e*) *S. officinalis*, respiration (r) followed by series of jet pulses (j) saturating recorder, and inhalant pulses negative to ambient pressure (i) (M. R. Clarke & E. R. Trueman, unpublished). (*f*) *Aeshna cyanea* larva, recording of pressure during respiration (r) and jet swimming (j) (from Hughes & Mill, 1966).

the end of recovery which was increased by about 0.02 m s^{-1} during the power stroke when water equal to some 25% of the mass of the medusa and retained water in the subumbrellar cavity was expelled.

The cubomedusan *Carybdea alata* resembles *Polyorchis* in manner of swimming and it proved possible to record pressure (E. R. Trueman, unpublished) and jet thrust with the medusa attached to a force transducer by a thread attached mid-dorsally which bore a small weight to keep the thread taut (Fig. 4a). A relatively low pressure of 20 Pa is generated during a power stroke of 0.3 s duration which increases the swimming velocity by about 0.02 m s^{-1}, the same value as for *Polyorchis*.

Bivalved molluscs

Infaunal bivalves, e.g. *Ensis*, exploit a specialised method of jet propulsion for burial (Trueman, 1975). Adduction of the valves expels water into the sand from the mantle cavity with pressures of up to 1 kPa. A similar method of producing a jet of water is also used in swimming by *Pecten* and *Chlamys*. Single pressure pulses of 0.6 kPa occur in *C. opercularis* but rapidly repeated adductions lead to a build up of pressure (Fig. 4d) (Moore & Trueman, 1971). Scallops expel water equivalent to as much as 50% of their body volume during each jet cycle and the use of this relatively large volume of water at low pressure suggests that scallops swim in an economical manner as respects energy consumption.

Dragonfly larvae

Jet propulsion in insects is unique to dragonfly larvae. *Aeshna cyanea* ejects water from a specialised rectal chamber by contractions of both dorso-ventral and longitudinal abdominal muscles at pressures up to 6 kPa at a frequency of 2.2 Hz (Fig. 4f) (Hughes & Mill, 1966; Mill & Pickard, 1975). This jet performance (Table 1) compares well with cephalopods, particularly when size is taken into account, although the latter may have a greater capacity for sustained swimming.

Salps

Rhythmic contractions of muscle bands in the body wall cause jets of water from the exhalant opening. These contractions are associated with the opening and closing of the inhalant and exhalant valves (Mackie & Bone, 1977). The inhalant aperture is closed some 40 ms before the body wall muscles contract to expel water posteriorly. Gentle stimulation of the inhal-

Table 1. Table summarising the performance of some animals swimming by jet propulsion; the wet body weight (or body weight plus contained water) is given when it is known, pressure refers to the range of pressures used in swimming where known and velocity to the maximum instantaneous velocity reported during a single jet cycle

Animal	Percentage of body volume expelled	Pressure (kPa) resp–max jet	Thrust (N)	Frequency of jet pulses (Hz)	Velocity (m s⁻¹)	Authority
Coelenterata						
Polyorchis, montereyensis	20–28	—	—	0.8	0.02	Gladfelter (1972a)
Carybdea alata 130 g	18	0.02	0.02	0.8	0.02	Trueman (unpublished)
Mollusca						
Notarchus punctatus	—	—	—	0.3	0.08	Martin (1966)
Chlamys opercularis	50	0.3–0.6	0.6	3	0.25	Moore & Trueman (1971)
Loligo vulgaris 350 g	50	0.5–30	5.0	2	2	Clarke & Trueman (unpublished)
Sepia officinalis 250 g	25	0.2–20	3.5	3	0.8	Packard (1969)
Octopus vulgaris 220 g	15	0.02–20	1.0	—	—	Trueman & Packard (1968)
Insecta						
Aeshna larvae	—	0.025–6	0.015	2–3.5	0.1	Hughes & Mill (1966)
Protochordata						
Salpa cylindrica	—	—	—	2	0.07	Madin (1974)

ant valve was shown by these authors to cause a break in the forward swimming rhythm and if the stimulus was strong enough, to cause reversal. This requires the exhalant aperture, now functionally the inhalant, to be closed before contraction of the body muscle bands. Bone & Mackie (1977) have made similar observations on *Doliolum* but we have no information on swimming velocity, jet pressure or thrust.

The recovery stroke

Fluid-filled cavities function in many invertebrates as a hydrostatic or fluid skeleton for the antagonism of muscle systems (Trueman, 1975). The cavity must function at a constant volume to allow one set of muscles to extend after contraction by force developed by their antagonist. Use of a fluid-filled cavity for jet propulsion implies a rapid change in the volume contained and clearly requires another mechanism to restore the fluid during the recovery stroke as well as to extend the contracted muscles. This is accomplished either by a special muscle system or by the storage of elastic energy, commonly in the integument.

Muscular mechanism

The mantle of jet-swimming coleoid cephalopods consists principally of circular muscle fibres interspersed by groups of radial muscle fibres and collagen tunics arranged with fibres in a trellis-like pattern (Fig. 2) (Ward & Wainwright, 1972). Packard & Trueman (1974) have demonstrated how the circular fibres deliver the power stroke causing the mantle to thicken and that the radial muscles contract so thinning the mantle, extending the circular muscle fibres and drawing water into the mantle cavity. One of the problems with this system is the possibility of the mantle wall folding as it expands and it is thought likely that the collagen trellis and some small residual tension in the circular fibres may prevent this. While expansion of the mantle involves contraction of the radial muscles, the elastic properties both of the muscle fibres and of the collagen trellis may well also play an important role. Another factor in respect of inhalation recently noted (M. R. Clarke and E. R. Trueman, unpublished) is the beating of the membranes at the base of the funnel which, when relaxed, cover the inhalant apertures. In *Sepia* these beat in the regular respiratory rhythm and are probably of prime importance during respiratory inhalation although of less importance during jetting.

During inhalation the mantle cavity of *Sepia* shows pressures of as much as 1.5 kPa below ambient for about 50 ms duration when the animal jets at high pressure and rapid frequency (Fig. 4e). This may well be caused by the contraction of the radial muscles of the mantle. It is possible to calculate a theoretical inhalant pressure provided the area of the jet and inhalant apertures are known. This is difficult to ascertain, but from analysis of cine film of *Sepia* taken during jetting the ratio of inhalant:funnel apertures is about 20:1. Assuming the duration of the power stroke and inhalation are equal then for the same volume of water moving in or out of the mantle cavity the ratio of inhalant to funnel velocities (q) will be 1:20. But the pressure is below ambient for only 50 ms, or about 1/5th the duration of the jet pulse. Thus the ratio becomes 1:4. The fluid velocity (q) is related to the pressure (P) generated as in the expression $q = \sqrt{(2P/\rho)}$, where ρ is the density of water. This may be derived from the Bernoulli equation if we assume losses at the apertures are negligible (Hunsaker & Rightmire, 1947). Thus an inhalant to jet pressure ratio of 1:16 is likely and for jet pulses of 20 kPa we might expect inhalant pressures of the order of 1.2 kPa below ambient. This corresponds reasonably well with the recordings made but it should be remembered that these values depend on the areas of the apertures used in jetting, these may be changed during swimming by the animal, and have not been measured accurately. It may also be noted that a lowering of jet frequency, with consequent protractions of the recovery phase, considerably reduces the pressure observed beneath ambient.

Elastic mechanisms

In all elastic mechanisms part of the energy of the power stroke is stored as elastic storage energy to be utilised in recovery of the contractile system with inhalation of water. This is clearly observed in scallops where part of the energy of the adductor muscles is stored in the rubber-like hinge ligament to cause the valves to open when the muscle relaxes. Investigations of its mechanical properties have shown it more efficient than the ligament of other bivalves (Trueman, 1953; Alexander, 1966) thus the repeated flapping of the valves in swimming is accomplished with minimum energy loss.

There is less known about the biomechanical properties of other elastic systems but Gladfelter (1972a) has shown how the mesoglea of *Polyorchis* is bent about a series of eight adradial structural joints during the power stroke whilst its recovery establishes the more circular shape of the medusa. In salps a similar system obtains where the integument antagonises the muscle bands of the body wall. Pickard & Mill (1974) have shown that cuticular deformation occurs in the anal region of dragonflies during respiration and it would seem likely that cuticular restoring forces would assist a single abdominal muscle, and possibly the diaphragm, during inhalation (Mill & Pickard, 1975). Pressures below ambient of about 1/3rd amplitude of the positive pulse have been demonstrated during normal respiration in dragonfly larvae (Hughes & Mill, 1966).

Discussion

Jet swimming has undoubtedly evolved independently in a number of invertebrate phyla and the performance of selected examples is presented in Table 1. Distinction should be drawn between cephalopods and salps with separate inhalant and jet apertures and animals with single apertures, such as jellyfish and *Notarchus*. It is common for a single aperture to be constricted at jetting, dilated at inhalation, e.g. dragonfly larvae (Mill & Pickard, 1972), *Notarchus* (Martin, 1966). This undoubtedly facilitates inhalation and increases the effectiveness of the jet pulse by increasing its duration. The latter would reduce the maximal swimming velocity but by so doing it would also reduce the resulting drag and the animal could travel further (Johnson *et al.*, 1972).

Many animals, such as squid, jellyfish and salps, are capable of prolonged swimming at a low and presumably economical speed. There is little information currently available on the cost of locomotion by jetting either from studies of comparative respirometry or by analysis of the hydromechanics of swimming. Packard (1969) has shown by stroboscopic photography that a $100\,g$ *Loligo* may accelerate at about $3.3\,g$ from rest to about $2\,m\,s^{-1}$ with a single jet cycle but this velocity is of short duration (*ca* $0.1\,s$) as the squid glides to a halt. His measurements suggest a high drag coefficient. High velocity may only be sustained by frequent powerful jet cycles and after producing these for some $5-30\,s$, experimental animals are commonly exhausted and may need an hour or more for recovery. These observations suggest that the cost of fast jet swimming is high.

Alexander (1978) has compared the Froude efficiency [useful power/(useful power + power lost to fluid)] of squid jet propulsion with trout of the same size ($220\,g$). A squid of this size would have a mantle capacity of about $100\,cm^3$ and would probably emit jets at a frequency of about $2\,s^{-1}$ so that it would impart backward momentum to about $200\,cm^3$ water s^{-1}. At a speed of $0.5\,m\,s^{-1}$ the trout gives backward momentum to $1400\,cm^3$ water s^{-1}. Thus at approximately similar mean swimming speeds the squid accelerates only 1/7th of the amount of water, but to a higher velocity than the trout, to produce the same forward thrust. Accordingly the

Froude efficiency of squid must be lower during fast swimming. We have as yet no data for the squid swimming at much lower speeds by jetting more gently and by use of its fins. But it is likely that sustained jet swimming at lower velocities will prove to be a much more economical means of locomotion. In rapid escape movements avoidance of a predator is more important than energy cost.

I thank Dr M. R. Clarke for reading the manuscript and, together with Dr A. Packard, for permission to incorporate unpublished data in this article.

References

Alexander, R. McN. (1966). Rubber-like properties of the inner hinge-ligament of the Pectinidae. *J. exp. Biol.* **44**, 119–30.

Alexander, R. McN. (1978). Swimming. In *Mechanics and Energetics of Animal Locomotion*, ed. R. McN. Alexander & G. Goldspink, pp. 222–48. London: Chapman and Hall.

Bone, Q. & Mackie, G. O. (1977). Ciliary arrest potentials, locomotion and skin impulses in *Doliolum* (Tunicata: Thaliacea). *Riv. Biol. Norm. Patol.* **3**, 181–91.

Gladfelter, W. B. (1972a). Structure and function of the locomotory system of *Polyorchis montereyensis* (Cnidaria, Hydrozoa). *Helgoländer wiss. Meeresunters.* **23**, 38–79.

Gladfelter, W. B. (1972b). Structure and function of the locomotory system of the scyphomedusa *Cyanea capillata*. *Mar. Biol.* **14**, 150–60.

Hughes, G. M. & Mill, P. J. (1966). Patterns of ventilation in dragonfly larvae. *J. exp. Biol.* **44**, 317–33.

Hunsaker, J. C. & Rightmire, B. G. (1947). *Engineering Applications of Fluid Mechanics*. New York: McGraw-Hill.

Johnson, W., Soden, P. D. & Trueman, E. R. (1972). A study in jet propulsion: an analysis of the motion of the squid, *Loligo vulgaris*. *J. exp. Biol.* **56**, 155–65.

Lagler, K. R., Bardach, J. E. & Miller R. R. (1962). *Ichthyology*. New York: Wiley.

Mackie, G. O. & Bone, Q. (1977). Locomotion and propagated skin impulses in salps (Tunicata: Thaliacea). *Biol. Bull. mar. biol. Lab. Woods Hole,* **153**, 180–97.

Madin, L. P. (1974). Field Studies of the biology of salps. Ph.D. Thesis, University of California, at Davis.

Martin, R. (1966). On the swimming behaviour and biology of *Notarchus punctatus* Phillipi (Gastropoda, Opisthobranchia). *Pubblicazioni Stn. Zool. Napoli*, **35**, 61–75.

Mill, P. J. & Pickard, R. S. (1972). Anal valve movement and normal ventilation in Aeshnid dragonfly larvae. *J. exp. Biol.* **56**, 537–43.

Mill, P. J. & Pickard, R. S. (1975). Jet propulsion in anisopteran dragonfly larvae. *J. comp. Physiol. A,* **97**. 329–38.

Moore, J. D. & Trueman, E. R. (1971). Swimming of the scallop. *Chlamys opercularis* (L) *J. exp. mar. Biol. Ecol.* **6**, 179–85.

Packard, A. (1969). Jet propulsion and the giant fibre response of *Loligo. Nature, Lond.* **221**, 875–7.

Packard, A. & Trueman, E. R. (1974). Muscular activity of the mantle of *Sepia* and *Loligo* (Cephalopoda) during respiration and jetting and its physiological interpretation. *J. exp. Biol.* **61**, 411–19.

Pickard, R. S. & Mill, P. J. (1974). Ventilatory movements of the abdomen and branchial apparatus in dragonfly larvae (Odonata: Anisoptera). *J. Zool., Lond.* **174**, 23–40.

Prandlt, L. & Tietgens, O. G. (1957). *Fundamentals of Hydro- and Aeromechanics*. New York: Dover.

Trueman, E. R. (1953). Observations on certain mechanical properties of the ligament of *Pecten. J. exp. Biol.* **30**, 453–67.

Trueman, E. R. (1975). *The Locomotion of Soft-bodied Animals*. London: Edward Arnold.

Trueman, E. R. & Packard, A. (1968). Motor performances of some cephalopods. *J. exp. Biol.* **49**, 495–507.

Ward, D. F. & Wainwright, S. A. (1972). Locomotory aspects of squid mantle structure. *J. Zool., Lond.* **167**, 437–49.

Weihs, D. (1977). Periodic jet propulsion of aquatic creatures. *Fortschr. Zool.* **24**, 171–5.

Young, J. Z. (1939). Fused neurones and synaptic contacts in the giant nerve fibres of Cephalopods. *Phil. Trans. R. Soc., Ser. B*, **229**, 465–503.

W. NACHTIGALL

Mechanics of swimming in water-beetles*

Water-beetles swim by using their hindlegs as underwater paddles to produce thrust. In contrast to the swimming of worms and fishes, the rigid trunks of water-beetles do not play any role in thrust production. On the other hand, the trunk, with its sub-elytral air bubble, does produce the necessary lift to prevent the animal from sinking. Since there is almost no hydrodynamic interaction between the thrust-generating mechanism of the hindlegs and the drag-generating trunk, these two systems can be studied separately; this is a great advantage and makes water-beetles ideal for the analysis of functional adaptation and mechanics of underwater movements. In this article I refer to my own investigations, which were carried out in two parts, 1960 to 1965 and in 1978. Although investigations are still in progress, so that many details cannot be discussed at the present time, some of the more recent, not yet published results are included in this paper and some of the questions under research are indicated.

Hydrodynamics of side-wheels as a technical analogue

In the world of technical machinery old-fashioned side-wheel paddle steamers provide a good analogue to find out what terms may be adequate in describing the biological system functionally. Fig. 1 demonstrates, that, to achieve a good degree of hydrodynamic efficiency, frontal areas and drag coefficients must have small values in a beetle's trunk and high values in its rowing apparatus during power stroke (see Hertel, 1963). How are these hydrodynamic necessities accomplished by nature?

* Financial support was given by Deutsche Forschungsgemeinschaft.

Drag-generating systems (trunks) in water-beetles

The drag of a body moved in a fluid is due to effects of inertia (flow separation, formation of eddies → inertial drag component) and viscosity (friction between sheets of the fluid → frictional component of drag). Using hydrodynamic balances one always measures total drag, which is the sum of these two components. Which of them is relatively greater depends on the body form and Reynolds number, but this question cannot be solved by direct measurement.

Forms and shapes

Fig. 2 shows photographs of four large species of the genus *Dytiscus*, the name-giving genus of the family of 'true water-beetles' (Dytiscidae). All of them are dorso-ventrally flattened and bear more or less sharp lateral

Fig. 1. Hydromechanical efficiency in a side-wheel paddle boat.

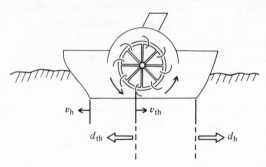

EXPLANATIONS:

a, drag area

$$p_{th} = d_{th} \, v_{th}; \quad p_h = d_h \, v_h;$$

$$\eta = \frac{p_h}{p_{th}} = \frac{d_h \, v_h}{d_{th} \, v_{th}};$$

For v_h = const. it follows that $|d_{th}| = |d_h|$, therefore

$$\eta = \frac{v_h}{v_{th}};$$

$$d_{th,h} = c_{w_{th,h}} \, a_{th,h} \, \frac{\rho}{2} \, (v_{th,h})^2;$$

solve for $v_{th,h}$ and transferred to $\eta = \dfrac{v_h}{v_{th}}$:

$$\eta = \frac{1}{1 + \sqrt{\dfrac{c_{d_h} \, a_h}{c_{d_{th}} \, a_{th}}}}$$

c_d, drag coefficient

d, drag

h, drag-generating system (hull of the ship)

th, thrust-generating system (side-wheel)

p, power

v, velocity

η, degree of efficiency

ρ, density of water

Fig. 2. Trunks of four species of the genus *Dytiscus* and geometrical bodies of same drag.

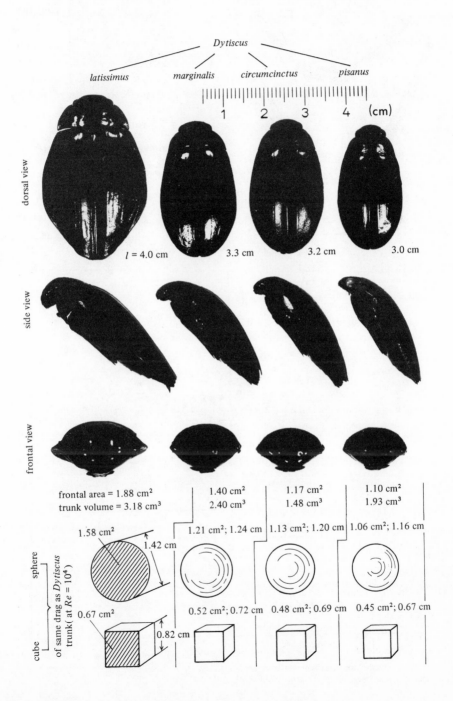

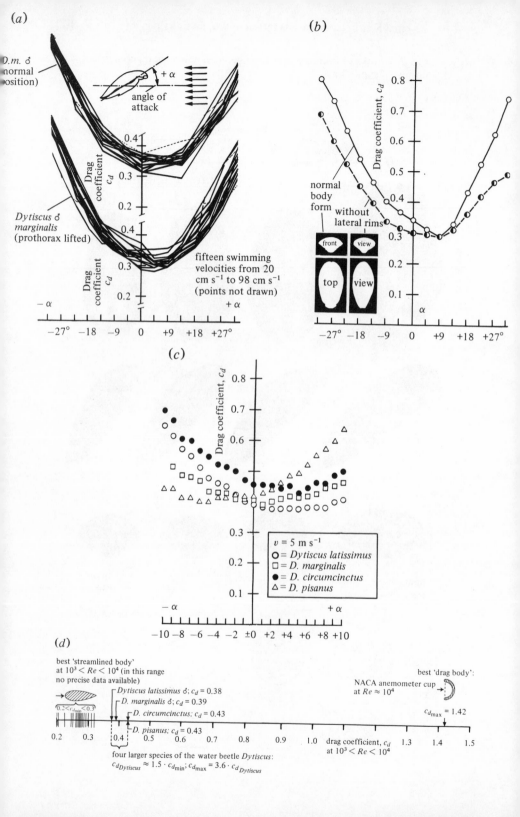

(a)

D.m. ♂ (normal position)

angle of attack   + α

Drag coefficient c_d

Dytiscus ♂ marginalis (prothorax lifted)

Drag coefficient c_d

fifteen swimming velocities from 20 cm s⁻¹ to 98 cm s⁻¹ (points not drawn)

− α   + α

−27° −18 −9   0 +9 +18 +27°

(b)

Drag coefficient, c_d

normal body form   without lateral rims

front   view

top   view

α

−27° −18 −9 0 +9 +18 +27°

(c)

Drag coefficient, c_d

$v = 5$ m s⁻¹
O = Dytiscus latissimus
□ = D. marginalis
● = D. circumcinctus
△ = D. pisanus

− α   + α

−10 −8 −6 −4 −2 ±0 +2 +4 +6 +8 +10

(d)

best 'streamlined body' at $10^3 < Re < 10^4$ (in this range no precise data available)

best 'drag body':

NACA anemometer cup at $Re \approx 10^4$

$0.2 < c_{d_{min}} < 0.3$

$c_{d_{max}} = 1.42$

Dytiscus latissimus ♂; $c_d = 0.38$
D. marginalis ♂; $c_d = 0.39$
D. circumcinctus; $c_d = 0.43$
D. pisanus; $c_d = 0.43$

0.2   0.3   0.4   0.5   0.6   0.7   0.8   0.9   1.0   drag coefficient, c_d 1.3 1.4 1.5
                                                      at $10^3 < Re < 10^4$

four larger species of the water beetle Dytiscus:
$c_{d_{Dytiscus}} \approx 1.5 \cdot c_{d_{min}}$; $c_{d_{max}} = 3.6 \cdot c_{d_{Dytiscus}}$

rims, i.e. prothoracic and elytral margins. All trunks appear smooth, 'stream-lined' and well adapted to produce the least possible drag. The pictures show front, dorsal and lateral views. The ventral surface appears to be stream-lined, too; front legs are folded into niches in the trunk during fast swimming, thus forming a smooth ventral plane. Now the question arises, how good are these body shapes hydrodynamically?

Frontal areas

The frontal area a, which is necessary to calculate drag coefficients, is obtained by photographing the beetle exactly parallel to the longitudinal axis using a telephoto lens (frontal view, see Fig. 2). The values of these areas, together with the values of the trunk volumes are indicated. Spheres and cubes of identical volume are drawn in, too. Frontal areas are relatively small, so that drag coefficients are expected to be small, too.

Coefficients of drag

Fig. 3c shows the total drag coefficients c_d of the four trunks shown in Fig. 2 as a function of the angle of attack α. This is the angle between the longitudinal axis of the body (line from upper rim of the eyes to abdominal tip) and the direction of flow. All measurements have been taken in a wind tunnel, using a semi-automatic balance-and-turning system (for details see Nachtigall, 1977c). Velocity was 5.0 m s^{-1} in air, corresponding to approximately 33 cm s^{-1} in water. This means fast-swimming velocity as found in larger beetles, for fractions of a minute, during hunting or escape reactions if disturbed. As was expected, c_d was minimal near $\alpha \approx 0$ (Fig. 3c). Absolute values are between 0.38 and 0.43, at Reynolds numbers $Re = vlv^{-1}$ between 9.9×10^3 and 13.2×10^4 (3 cm $< l <$ 4 cm; $v_{water} = 33$ cm s^{-1} = const; $v \approx 1 \times 10^{-6}$ m^2 s^{-1}; compare Nachtigall, 1977b for a general discussion of biological aspects of Reynolds number). In this range, technical objects show c_d-values between approximately $0.2 \div 0.3$ (best low-drag form) and 1.42 (best high-drag form; anemometer cup) (Fig. 3d). The values coincide fairly well with earlier drag measurements in Dytiscus marginalis ($c_d = 0.30 \div 0.38$ at $\alpha = 0°$) made by my co-worker D. Bilo and myself in 1965 (Fig. 3a). (The fact

Fig. 3. (a) Total drag coefficients of Dytiscus marginalis as a function of the angle of attack. (b) Lateral rims of Dytiscus marginalis and their influence on drag. (c) Total drag coefficients of four species of Dytiscus. (d) Dytiscus trunks and models of maximal and minimal drag coefficient.

that they are somewhat higher may be due to small deviations of the hot-wire instrument used. Preliminary calibrations show, that c_d values could be 13% smaller, maximally. Calibration is still in progress.) Therefore, *Dytiscus* trunks are rather well-adapted low-drag bodies, but do not reach the best possible configuration. This is mainly due to the shape of the lateral edges (Fig. 3*b*) which are necessary to stabilise swimming against rotational oscillation around the *y* and *z* axes. Models of lower c_d-values are unstable without trailing edge fins; after small deviations they quickly turn obliquely against the flow. This shows that dytiscid bodies are technically speaking 'compromise constructions'; they combine relatively low drag with good stability of motion. An analysis of the volume drag coefficients of water-beetle bodies, compared with geometric shapes, is being made at the present time in my lab. Data are not yet available, but we expect that, with respect to this function, the bodies are optimal, too. Another question is the dependence of drag coefficients on Reynolds number. Technical measurements lead one to expect that smaller, slower swimming beetles should show higher c_d-values (Fig. 4). This means, that they should need relatively more energy for swimming. A scale effect analysis of the energetics and hydrodynamical adaptations of various sizes of dytiscids was made Nachtigall (1977*a*). Due to a lack of data all conclusions remain hypothetical at the present time.

Fig. 4. Expected changes of drag coefficients in smaller, slower swimming dytiscids (dotted line), derived from technical measurements.

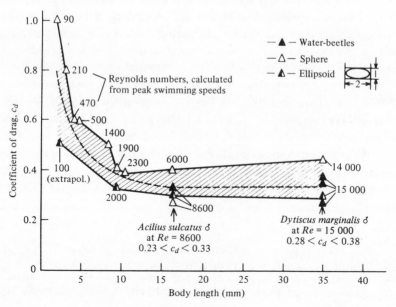

Pressure distribution

Drag adaptation can also be calculated from measurements of pressure distribution. Together with total force measurements it may be possible one day to distinguish between inertial and frictional drag components. As direct pressure measurements on the small original trunks seem to be (almost) impossible, we built models in ten-fold magnification (Nachtigall, Rothe & Rothe, 1978a, b). The latter contain up to 166 static pressure holes. In a first experimental set up the pressure distribution was analysed as a function of the angle of attack in a higher Reynolds number range ($4 < v < 22$ m s^{-1}; $9.3 \times 10^4 < Re < 5.2 \times 10^5$). This technical method may help to answer the question whether such unconventional forms like the flattened bodies of dytiscid beetles may be used as optimal forms for technical purposes, (e.g. for small submarines). Fig. 5 shows one of the first results. There is positive pressure in the head region of the dorsal (suction) side, followed by negative pressure. This reminds one of the pressure distribution of an aerofoil of small aspect ratio. Quantitive and qualitative changes in pressure distribution due to Reynolds number (Fig. 5) and the angle of attack (not shown) could be detected. A critical discussion of these data and phenomena will be given elsewhere. Measurements of pressure distribution in the models at Reynold's number of the original beetle are difficult to make due to the very small pressure differences. The special equipment required is under construction.

Thrust-generating mechanisms (rowing legs) in water-beetles

It was shown that, for the trunks of water-beetles, the principles presented in Fig. 1 are fulfilled: frontal areas, a are small, drag coefficients, c_d, are small. The contrary should be the case in the rowing mechanism.

Form and shape of rowing legs

In Fig. 6 the rowing legs of the medium-sized water-beetle *Acilius sulcatus* (Dytiscidae) and of the small whirling beetle *Gyrinus natator* (Gyrinidae) are shown together with details of the rowing appendages, i.e. hairs of blades. All parts of the leg – tibia and the tarsal parts – are broadly flattened. The surface of the rowing appendages is greatly enlarged. These legs beat synchronously until they hit each other in the medium sagittal plane (Fig. 7). During the power stroke they are stretched and move with the flattened side, enlarged by the rowing appendages, which are maximally stretched, too (Figs. 8 and 10), perpendicularly against the fluid. So their

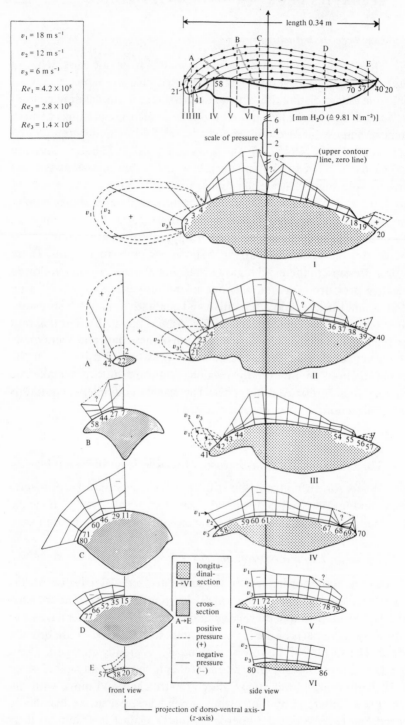

Pressure distribution of *Dytiscus marginalis* ♂

Fig. 5. Example of a measurement of pressure distribution around a model trunk of a male *Dytiscus marginalis*.

Fig. 6. Scanning electron micrographs of the hindlegs and rowing appendages of *Acilius sulcatus* (*a*, *b*) and *Gyrinus natator* (*c*, *d*, *e*).

frontal area *a* is as large as possible. The flattened parts must be character-ised by high c_d-values around 1, like flat discs, moving perpendicularly through the fluid. Therefore, the demands of the equation in Fig. 1 seem to be fulfilled, too, especially in the *Gyrinus* hindleg. It forms, together with its overlapping rowing blades, a comparatively large, oval blade of light camber with very sharp edges, which is moved with the concave side against the water (Fig. 10). In rowing hairs there are certain complications.

Rowing hairs

A rowing hair of *Dytiscus* is maximally 4 mm long and from 20 to 2 μm in diameter from base to tip, showing a length/diameter ratio of 200:1 to 2000:1. It is a remarkable structure in at least three ways. Firstly, hairs are

Fig. 7. Beating movement of the hindleg of *Acilius sulcatus* during a stroke cycle.

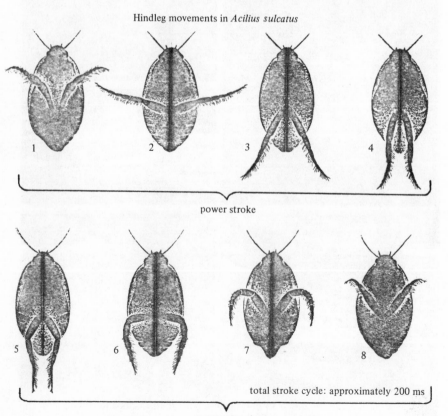

Hindleg movements in *Acilius sulcatus*

power stroke

total stroke cycle: approximately 200 ms

recovery stroke

inserted with a special gearing mechanism at the edge of the tibia or tarsus in such a manner, that they are very easily spread by water pressure at the beginning of the power stroke (Fig. 8) until they hit a common ridge (Fig. 6c), thus forming an accessory propulsion area. At the beginning of recovery stroke they are laid against the leg automatically by water pressure, so that

Fig. 8. High-speed frames (220 frames per second) of the power stroke of *Acilius*, viewed from behind. First quarter (turning of the leg in power stroke position) not shown. Swimming hairs are turned to working position within less than 1/40 s.

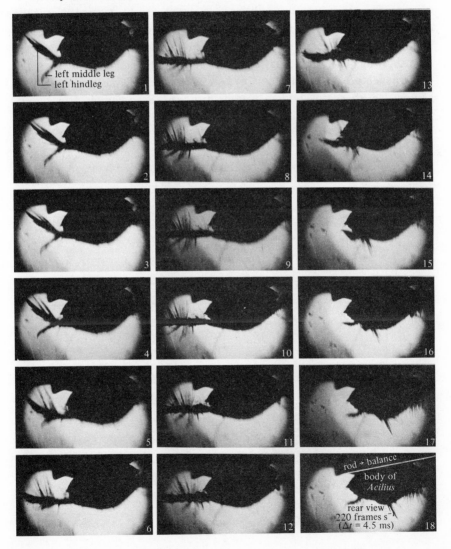

the accessory area now completely vanishes. There is a kind of a snapping apparatus at the bases of the swimming hairs, keeping them in spread or folded position (Fig. 6b). The same is true in the rowing blades of *Gyrinus* (Fig. 6c–d) (Nachtigall, 1961). Secondly, due to their arrangement in the form of a 'hydromechanical rake', they generate rather high drag with minimal material, i.e. minimal frontal area. It was shown by Nachtigall (1960), that 68% of the drag of an outstretched *Acilius* leg is due to the swimming hairs, which comprise 80% of the total area of the leg in power stroke position. It is my hypothesis, that the distance–diameter relation of the rake of swimming hairs is optimised. Measurements to solve this question are in progress. Thirdly, neither the hairs nor the blades bend, let alone crack, even during very fast beats. They show remarkable stretching and bending characteristics which seem to be due to a two component construction, showing different mechanical peculiarities of the inner cylinder and the outer envelope. Material seems to be arranged in such a way as to form geometrical shapes of 'bodies of same firmness'. Scanning electron microscope investigations, together with bending experiments, are in progress.

Fig. 9. Power stroke (strobe illumination, *a*) and sketch of hydromechanical force components (b) in left hindleg of *Acilius sulcatus*; (ventral view).

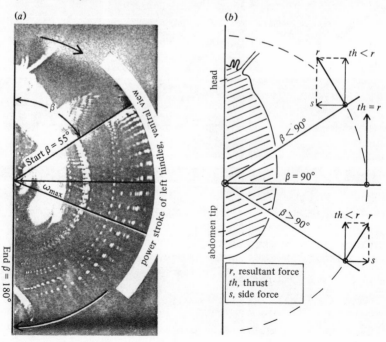

(a)

(b)

Power stroke and hydromechanical force components

Fig. 9*a* shows a multiple flash photograph of the power stroke of *Acilius sulcatus*. A series of white dots on the left hindleg is illuminated by a stroboscope of 220 flashes s^{-1}. It is shown that the stroke angle β goes from 55° to approximately 180° and the angular velocity $\omega = l\,v$ (l = distance between a point on the leg from fulcrum, v = absolute velocity of this point) is maximal at 90° < β < 110°. Here thrust *th* is identical with the resultant hydrodynamic force *r*. As *th* = *r* sin β the beating leg induces a side force component *s* before (β < 90°) and after (β > 90°) reaching the position perpendicular to the median axis. The side force is directed to the median line in the first half (β < 90°), against it in the second half (β > 90°). In straight ahead swimming, the two side force components of the two synchronously beating legs add up to 0, thus costing energy without inducing thrust, and reducing the hydrodynamic degree of efficiency of the power stroke. As $r \sim \omega^2$ it is especially important that r_{max} is turned to thrust with minimal loss. This is exactly the case in the *Acilius* larvae (Nachtigall, 1962) when ω_{max} is achieved at $\beta = 90°$, so that $s = 0$ at $r = r_{max}$. In the imago of *Acilius* r_{max} is turned to thrust at a rate of 100 to 90% (Nachtigall, 1960), as ω_{max} is achieved at 90° < β < 110°. In *Gyrinus* there is a broad maximum of ω at 90° < β < 130° (Nachtigall, 1961).

Change of shape of the rowing leg during the stroke cycle and hydrodynamic efficiency

The underwater paddles of the water-beetles induce a forward movement of the body if the net thrust, i.e. the thrust of the power stroke minus the counter-thrust of the recovery stroke, is greater than zero. There are two possibilities to fulfil this condition. Firstly, power stroke velocity may be greater than recovery stroke velocity, whereby the shape of the leg remains unchanged during the whole stroke cycle. Secondly, the thrust-generating area in power stroke may be greater than the counter-thrust-generating area in recovery stroke, whereby velocity distribution remains the same during the two stroke phases. Nature nearly always relies so totally on the second principle, that it can even afford to make the recovery stroke slightly faster than the power stroke, thus enhancing the frequency of beating which again means more possibilities for course corrections in a given time.

How drastically shape and area of a rowing leg are changed during the stroke cycle is demonstrated best in the hindleg of the whirling beetle *Gyrinus* (Figs. 10 and 11a). During power stroke, all parts of the leg–femur (partly), tibia and the paper-thin flattened sections of tarsus – are maximally spread and face the fluid with their broad side; the fan of the rotating blades is spread. During recovery stroke the broad sides are turned up to 90° and drawn forward, sliding along the ventral side (probably wholly in the ventral side's own boundary layer). Tarsal sections are folded together like a set of playing cards and folded into a flat cavity of the femur. The area of the rowing blades vanishes totally, thus creating astonishingly different aspects in the frontal area of the leg during the two stroke phases (Fig. 10a, c). Some kinematic parameters are indicated in Fig. 11b, together with model

Fig. 10. Different aspects of a *Gyrinus* hindleg in the middle of power stroke (a, right leg, rear view) and recovery stroke (c, left leg, front view). (b) shows a partly folded leg in the middle of recovery stroke, ventral view. Muscles are indicated.

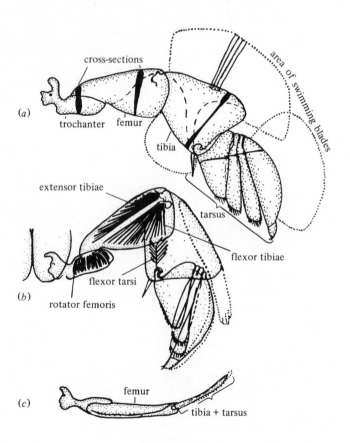

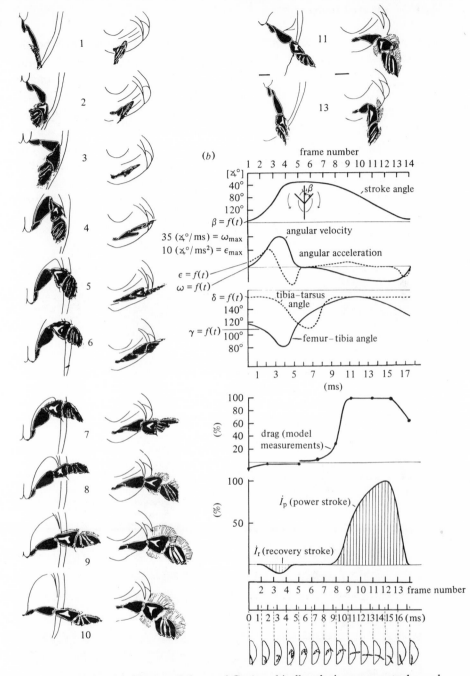

Fig. 11. (a) Change of shape of *Gyrinus* hindleg during power stroke and recovery stroke. (b) stroke angle, β, plotted against time; angular velocity, ω, plotted against time. From this and point-to-point drag measurements of a leg model, a plot of stroke impulse against time is constructed. It can be seen that power stroke impulse is much greater than recovery stroke impulse.

measurements of drag of different stroke positions. Altogether the change in power stroke impulse

$$\dot{I}_p \int_{t_1 = 7\,\mathrm{ms}}^{t_2 = 18\,\mathrm{ms}} v\,(t)\mathrm{d}t \qquad (v,\ \text{velocity})$$

is up to forty times greater than change in recovery stroke impulse

$$\dot{I}_r \int_{t'_1 = 0\,\mathrm{ms}}^{t'_2 = 7\,\mathrm{ms}} v\,(t)\mathrm{d}t$$

Fig. 12. Beating movements of *Dytiscus marginalis* ♂ swimming at 10.4 cm s⁻¹ (*a*) and 15.3 cm s⁻¹ (*b*). Base and tip of a hindleg connected by a straight line. (*c*) Hindleg of *Acilius sulcatus* to demonstrate relative areas of the single parts.

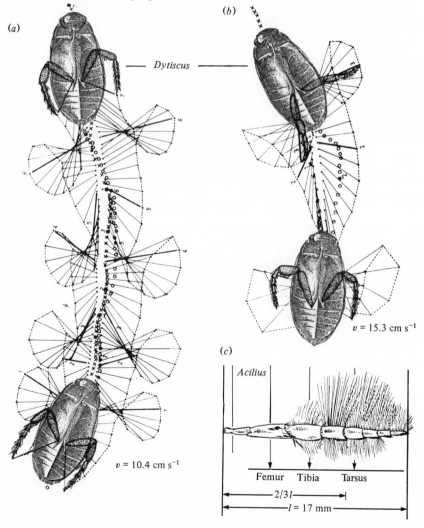

(*a*)

(*b*)

—— *Dytiscus* ——

v = 15.3 cm s⁻¹

v = 10.4 cm s⁻¹

(*c*)

Acilius

Femur   Tibia   Tarsus

2/3 *l*

l = 17 mm

providing a hydrodynamic degree of efficiency of $\eta = I_p/(I_p + I_r) = 0.87$, which is very high for a rowing system.

The position of the 'fixpoint' of a rowing leg is also a good indication of its hydrodynamic quality. This is the point on the longitudinal axis l of a technical paddle or rowing leg, that remains fixed in space, whereas, during power stroke, the parts of the paddles lying distally from this point, move backwards, the parts proximally from it move forwards. In good paddle boats, the 'fixed point' lies far outward, at three-quarters l or even more distally. Fig. 12 shows examples of a 3.3-cm *Dytiscus marginalis*, performing slow (*a*) and medium (*b*) beating movements. It is shown, that the 'fixed point', demonstrated by the main overlapping area, lies around one-third l to one-half l, nearer to one-third l. This means that an optimum in the technical sense of a paddle (characterised by a broad area near the end, which is taken out of the water during recovery stroke) cannot be the aim of the animal's thrust mechanism. But there is a remarkable coincidence between morphological configuration, kinematics of movement and force generation in these rowing legs. The 'fixed point' lies near the femur–tibia joint. All hairy parts of the leg are situated distally from it. Thus even the proximal swimming hairs on the tibia contribute to thrust. The greatest part of the area lies at two-thirds l to three-quarters l (Fig. 12*c*). This is where the maximum of the useful forces is generated in beating legs, oscillating wings and propellers in nature and machines. For more details see the summarising articles of Nachtigall (1964; 1973).

References

Hertel, H. (1963). *Biologie und Technik*. Mainz: Krausskopf.

Nachtigall, W. (1960). Über Kinematik, Dynamik und Energetik des Schwimmens einheimischer Dytisciden. *Z. vgl. Physiol*. **43**, 48–118.

Nachtigall, W. (1961). Funktionelle Morphologie, Kinematik und Hydromechanik des Ruderapparates von Gyrinus. *Z. vgl. Physiol*. **45**, 193–226.

Nachtigall, W. (1962). Die Mechanik der Schwimmbewegungen bei der Acilius-Larve (Coleoptera, Dytiscidae). *Int. Revue gesamten Hydrobiol*. **47**, 413–30.

Nachtigall, W. (1964). Die Schwimmechanik der Wasserinsekten. *Ergeb. Biol*. **27**, 39–78.

Nachtigall, W. (1973). Locomotion: mechanics and hydromechanics of swimming in aquatic insects. In *The Physiology of Insecta*, 2nd edn, vol. 3, ed. M. Rockstein. London & New York: Academic Press.

Nachtigall, W. (1977a). Swimming mechanics and energetics of locomotion of variously sized water beetles – Dytiscidae, body length 2 to 35 mm. In *Scale Effects in Animal Locomotion*, ed. T. J. Pedley. London & New York: Academic Press.

Nachtigall, W. (1977b). Zur Bedeutung der Reynoldszahl und der damit zusammenhängenden strömungsmechanischen Phänomene in der Schwimmphysiologie und Flugbiophysik. *Fortschr. Zool.* **24** (2/3), 13–56.

Nachtigall, W. (1977c). Die aerodynamische Polare des Tipula-Flügels und eine Einrichtung zur halbautomatischen Polarenaufnahme. *Fortschr. Zool.* **24** (2/3), 347–52.

Nachtigall, W. (1978). Statement: Evolutionsaspekte und physikalische Genzwerte. *Zool. Jahrb. Abt. Anat.* **99**, 109–14.

Nachtigall, W. & Bilo, D (1965). Die Strömungsmechanik des Dytiscus-Rumpfes. *Z. vgl. Physiol.* **50**, 371–401.

Nachtigall, W., Rothe, H. J. & Rothe, U. (1978a). Eine Einbettungsmethode zur Vermessung kleiner biologischer Objekte. *Präparator,* **2**, 233–5.

Nachtigall, W., Rothe, H. J. & Rothe, U. (1978b). Geometrisch exakte Modellkonstruktion der Rümpfe von Wasserkäfern. *Präparator,* **3**. 262–5.

C. S. WARDLE & J. J. VIDELER

Fish swimming

Introduction

There are more than 18 000 species of living teleosts and 500 species of elasmobranchs showing a great range of adaptation to particular ways of living in both marine and fresh waters. These adaptations include many variations on the use of body structure and fins for specialised methods of swimming. We are confining our attention here to the straightforward swimming of fish like cod, *Gadus morhua*; salmon, *Salmo salar* and mackerel, *Scomber scombrus*.

Casual observation of these fish shows that they swim by sweeping the tail fin from side to side. The dorsal and ventral edges of the tail blade always lead the sweep. The tail tip shows the greatest lateral amplitude of any movement seen on the fish's body. The tail apparently moves due to bends developed somewhere in the middle part of the body. Speed of swimming is related to the frequency of the tail sweeps. A fish can swim for long periods holding station in a water current or while moving over long distances. The same fish when startled can burst away at apparently high speed using rapid sweeps of the tail blade. Accessory fins and attitudes of the body can all be adjusted allowing small changes in position. The fish can steer, glide, turn rapidly or it can break suddenly and precisely by freezing a carefully shaped wave on its body and fins. Every species of fish shows, as well as these basic manoeuvres, some specialities such as fast starting in the cod, leaping in the salmon and high-speed cruising in the mackerel which often involve modification of the fish's swimming apparatus.

Dissection of any one of these fish will show that it is a vertebrate with the characteristic, incompressible but bendable spinal column. The obvious lateral swimming muscles consist of complicated shaped segments of muscle (myotomes) connected by sheets of collagen (myosepts). The bulk of the muscle is white and a thin layer of red muscle is found on the surface. The muscle is covered by a tough skin extending over the whole body and fins. The fish considered here all have a prominent tail blade as well as a variety of other fins.

Fish have gills to obtain oxygen from water, red blood cells are pumped by a heart through blood vessels carrying oxygen to the red muscle layers where contraction derives energy by burning a variety of fuel stores. The white muscle has a large store of glycogen which is converted without oxygen to lactic acid during fast swimming. The aerobic and anaerobic energy sources set limits to cruising speed and fast-swimming endurance. The different muscle types have evolved with limited shortening times which in turn limit swimming performance. Detailed knowledge of all these and other physiological features of the fish can be related to the observed swimming movements.

The present discussion will examine the relevant details of the movement, the anatomy and the physiology with the aim of pointing out relationships and also areas where more knowledge is required.

The anatomical and physiological mechanisms can only be fully understood when the movements have been completely described. Our understanding of fish swimming is still surprisingly limited because systematic analysis of even the straightforward steady swimming movements hardly exists. The main reason why we still know so little is because it is very difficult to make accurate films of swimming fish. An aquarium tank with static water seems to give the best chance to record the swimming movements.

Flume tanks where the water moves through a tube or channel are essential for example when determining how much oxygen is being used during swimming but the inevitable complications of water flow generated in flumes should be avoided when studying the swimming movements themselves. The size of the static tank should suit the size of the fish to be studied or it will limit the performance. The 10-m diameter annular tank with rotating gantry and static controlled water conditions was built at the Marine Laboratory Aberdeen in order to explore the swimming performance of larger commercially important fish. Even this tank has been found to inhibit the maximum swimming speed performance of larger cod (Wardle, 1975). The techniques of recording the swimming movements add further difficulties to this research. High frame rate cine-cameras require intense light levels (9 kW to illuminate 1 m^3; Videler & Wardle, 1978) which inevitably inhibit all but the most tolerant fish. The much more sensitive television cameras can be modified to give high frame rates (C. S. Wardle, unpublished) and strobe lights can freeze the images (Wardle, 1975) but the resulting image detail recorded on video-tape or disc is poorer and must be allowed for. Methods of catching, handling and training each species to be studied must be developed in order to preserve the performance and behaviour characteristics of the species in captivity. Bainbridge (1958) with his fish wheel was able to make valuable recordings of small fish swimming in static water conditions. The tanks at Aberdeen allow extension of the size range.

The techniques developed to control the swimming behaviour of the captive fish and record their movements are described in Wardle & Kanwisher (1974), Wardle (1975), Pitcher, Partridge & Wardle (1976), Wardle & Reid (1977) and Videler & Wardle (1978).

Swimming movements

Model of the two wave system

The movements of the body of a fish like cod during straightforward locomotion can be described as waves of lateral curvature that pass along the body from head to tail. Numerous authors have described fish swimming movements, but Gray (1933a) showed most clearly how these can be understood as a combination of two wave-like phenomena: 1. Sinusoidal tracks described in space by every single point of the body (forward speed u, wavelength λ_s, wave-period T_s and amplitude A. 2. Cyclic changes of the curved shape of the body itself, further referred to as 'the body wave' (backward speed v, wavelength λ_b, wave-period T_b).

We will use a simple model to investigate the basic relationships between the parameters of the two wave systems. The body of a fish in a horizontal plane could conveniently be represented by a single line. Our models in Fig. 1a–c show three different ways in which a wave of curvature could pass along a line. In Fig. 1 is indicated the mean direction of forward motion, the X-direction and that perpendicular in the horizontal plane, the Z-direction. Fig. 1a–c illustrates the whole range of possibilities for the case where the amplitude A (the excursion in the Z-direction) is equal for all parts of the line and assuming that there is a certain wavelength of the body wave λ_b and a certain body wave-period T_b where

$$v = \lambda_b/T_b$$

There is an equal time separation between adjacent lines and we choose the right end of two lines in each case to determine the velocity and direction of movement of this part of the line at a certain instant. As in the real fish we assume that the movements of the lines have propulsive effect and that the line encounters resistance. In Fig. 1a the propulsive effect of the movements of the line is counterbalanced by the resistance, and the forward speed is zero. Each track is an extreme case in which the resulting forward speed u and the mean value of the wavelength of the track of each part of the line, λ_s, are both zero.

$$u = \lambda_s/T_s$$

The body of a fish and the model lines are entireties which demand that the period of the body wave (T_b) is equal to the period of the wave-track in space for every single point of the body (T_s) (Videler & Wardle, 1978) so:

$$T_s = T_b = T$$

Confirmation of this statement is easily obtained from Fig. 1a. This figure shows also how a point of maximum curvature travels from the mid-region to the right end, indicating the wave speed v.

Fig. 1. Model showing the basic relationships between the parameters of two wave systems existing simultaneously on an inextensible line. (See text for further details.)

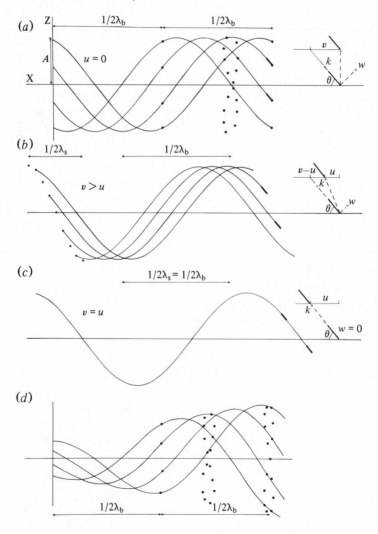

In contrast to Fig. 1*a*, Fig. 1*c* shows the extreme case in which the speed of the line *u* is equal to the speed of the body wave *v*. The first three equations show that in this case $\lambda_s = \lambda_b$ and the line is following one single path. The real swimming fish could conceivably copy this model line under two circumstances: In the 'imaginary' case where there would be no slip between waving body and water and in the 'realistic' case in which the speed of the fish decreases and the fish free-wheels its body wave by continuously keeping its wave speed *v* equal to forward speed *u*. Straightforward swimming in which a steady wave speed results in a steady forward speed, is illustrated in Fig. 1*b*. In this case the speed of the body wave is greater than the forward speed and consequently λ_b is greater than λ_s.

The main purpose of studying kinematic data of a swimming fish is to learn about the interaction between the surface of the fish and the surrounding water. Knowledge of the main wave parameters only gives a crude impression of this interaction. Additionally the speed and direction of movement relative to the water of each single point of the body should be known. In our simple model in Fig. 1*a*–*c* it is sufficient to regard the behaviour of a small part only. For convenience we have chosen the right end of the model line and look at it in each of the Figs. 1*a*–*c* at two successive instances. (In Fig. 1*a*, *b*, *c* the pieces under consideration are indicated by thick lines.) The distance moved between these two end positions allows calculations of instantaneous speed and measurement of the direction of displacement. The line end positions in each of the Figs. 1*a*, *b* and *c* are redrawn (to the right) in separate enlarged diagrams. These diagrams define angle θ as the angle between the body surface and the mean path of motion and show the direction and magnitude of the line end velocity, drawn as a dashed vector. The line end velocity vector is resolved into a component *k* tangential to the line end and a component *w* normal to the line end.

If the forward velocity *u* is very small compared to the wave velocity *v*, as in Fig. 1*a*, the dashed velocity vector will point in the Z-direction. The component *w* is relatively large and *k* is consequently small. In the other extreme case of Fig. 1*c* the velocity vector follows the direction of component *k* and there is no speed vector in the *w* direction. The difference between *v* and *u* and the value of angle θ determine the direction and magnitude of *k* and *w*. From the diagram (Fig. 1*b*):

$$w = (v - u) \sin \theta$$

and
$$k = v/\cos \theta - (v - u) \cos \theta$$

During the swimming cycle, θ varies in a sinusoidal way. Analysis of the two waves in these models has indicated a set of key parameters that can now be used to examine and define the same waves in the real swimming fish.

Relation between the model and the kinematics of a fish

We will use Fig. 2 to have a close look at the movements of the body of a fish. In this example a cod (0.42 m) was filmed during straightforward swimming at a steady speed (u = 0.7 m s^{-1}). Careful analysis of lateral displacements of a number of points of the body, in the manner described by Videler & Wardle (1978), provides knowledge about the exact position and speed of wave crests travelling down the body. These wave crests (identified by the dashed lines in Fig. 2), allow measurement of the wave speed v = 0.9 m s^{-1}, the magnitude of the wave-period T, the body wavelength λ_b and λ_s which is the distance moved forwards after one wave-period T.

Fig. 2. Drawings of the outline of a 0.42 m long cod, swimming steadily in a straightforward direction. The drawings were made from a cine film taken from the dorsal side at 200 frames s^{-1}. Every fourth frame of one tailsweep is shown. The time separation between the frames in the sequence from the left to the right is 0.02 s. The small circles indicate the same two fixed positions on the background, drawn with each frame. We can assume that a wave crest indicates a point on the body where muscles on one side are fully contracted and on the opposite side are fully relaxed. Note how there are normally two wave crests on the body at any time. The width of the hatched bars indicates how the relative amount of muscle fibres involved in the contraction might increase and decrease.

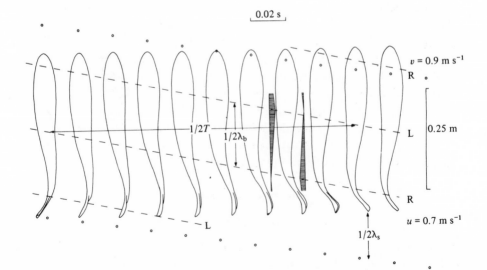

It is clear from Fig. 2 that the cod differs from the model Fig. 1b and always shows more than one complete wavelength on the body and there is a distinct increase in amplitude of the lateral movement towards the tail tip. The consequence of both these deviations from the simple model becomes clear by comparing Figs. 1a and 1d (Fig. 1a shows the model with constant amplitude from head to tail tip and Fig. 1d is drawn with a linear amplitude increase. The pattern of lateral displacement of a number of points on the body is indicated by the dots.) Note how two of these points at distances $\frac{1}{2}\lambda_b$ and $1\lambda_b$ from the left side of Fig. 1a move in nearly straight lines of displacement. All other points on the line describe a figure of 8 when the lines are superimposed as shown (Gray, 1933a). In the other model (Fig. 1d), while the point at exactly $\frac{1}{2}\lambda_b$ from the left side moves in a nearly straight line, the point at $1\lambda_b$ makes a narrow but recognisable figure of 8 movement. All other points in this model make larger figure of 8 movements. By measuring values for the parameters outlined here movements of the swimming fish recorded on film or video-tape can be accurately described and compared.

In Fig. 3 the swimming movements of the cod shown in Fig. 2 are compared with sine waves having the same dimensions. Two extreme and one intermediate positions are drawn for the centre line of the cod and these can be compared with three sine waves with the same values for amplitude of the wave along the body and the same body wavelength. The cod makes movements very similar to the sine wave but differs in the anterior region where the stiff head region does not show the small amplitude bends seen on the sine waves. In the last part of the tail there are small differences in detail that need further investigation. These findings contrast markedly with descriptions of vertebrate swimming movements given by Blight (1977) where the existence of a backwards moving wave was questioned.

Fig. 4 shows that swimming of a mackerel can also be analysed using the

Fig. 3. (a) Superimposed centre lines at three different instants during straightforward swimming at $0.7\,\mathrm{m\,s^{-1}}$ of a 0.42 m cod as in Fig. 2.
(b) Calculated lines based on a sine wave with λ_b and amplitudes (A) as in Fig. 3a, using the equation $z_{(x)} = A \sin 2\pi x/\lambda_b$. (Coordinates z and x are shown in the Fig.)

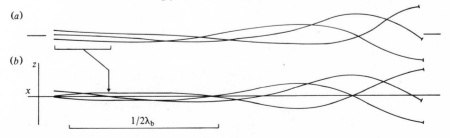

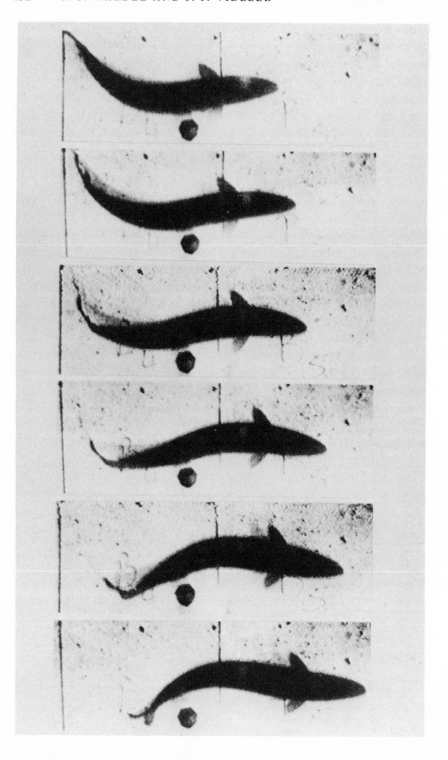

parameters of the two wave system. The number of pictures suitable for this type of analysis is very limited and the key to further understanding of fish swimming is a systematic collection of such pictures of swimming fish. The records should be made with a number of different species and include a range of sizes swimming at all speeds possible.

Movements and body structure

How are fish swimming movements made?

There are now repeated electromyographic observations indicating that slow swimming movements of teleost fish are made using the thin layer of red muscle which forms the outermost layer of muscle fibres on each of the myotomes (Rayner & Keenan, 1967; Bone, Kiceniuk & Jones, 1978). For example in cod these outer surfaces (Figs. 5 and 6) closely adhere to the underside of the skin and the fibres are in line with the longitudinal body axis. The ends of the muscle fibres of adjacent myotomes join firmly to each side of the myosept and the myosepts are fused into the skin. The thin layer of red fibres is divided into segments by the myosepts. The segmented structure permits any one or a number of the discrete segmental units to be contracted. This can allow a slow body wave as well as a variety of other slow movements to be accomplished.

Bends in the body are made by a number of segments contracting on one side while simultaneously segments in adjacent regions contract on the other side of the body. Between these opposing lateral muscles the vertebral column maintains a fixed length being incompressible. The vertebral column

---

Fig. 4. Mackerel, *Scomber scombrus*, weight 0.3815 kg, length 0.368 m, velocity 2.83 m s^{-1}, frames 0.010 s. The sequence shows one left-to-right sweep of the tail with λ_s or stride = 0.7L. Two zones of shortening of the lateral muscle are seen as bending particularly in the third frame where lateral muscles are shortening just behind the right pectoral fin and on the left side just ahead of the tail peduncle. The zones of bending move towards the tail in the subsequent frames.

The images were made using a television camera modified to scan a centre ½ section of the camera tube twice during the time for one normal full scan giving 100 images s^{-1} on video-tape. The silicon-diode array tube (25 mm) is red sensitive and has a short image retention time. A synchronised strobed (10 μs flash) red light source placed next to the camera lens illuminates frozen silhouettes of the mackerel against a screen of reflex-reflector material (3M Scotchlite 'High Intensity'). These reproduced frames are not suitable for detailed analysis due to the TV monitor distortions.

Fig. 5. Schematic drawing of the surface of a cod with the skin removed. The decrease of the span between the myosepts towards the tail is slightly exaggerated to show how myosepts become more in line with the longitudinal axis towards the tail. Note how the area covered by red muscle coincides with the zone where the skin is firmly attached to the myosept.

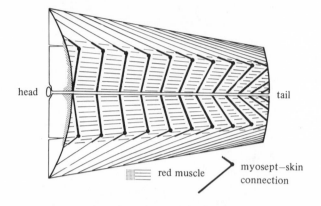

head ⊙   tail

red muscle   myosept–skin connection

Fig. 6. A semi-schematic drawing in oblique projection of one myosept of a cod. The myosept consists of a dorsal and a ventral sheet. The sheets coincide near the surface in the higher horizontal plane shown. The ventral sheet has a complicated 'boot' shape. The tip of the boot extends forward beyond the attachment to the vertical septum. Two shaded transverse sections show the adjacent myosepts and heavier shading indicates the area where the muscle tissue between the myosept is red instead of white.

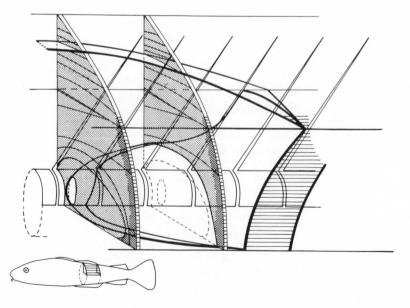

is easily bent in the horizontal plane and an elaborate system of cushions and elastic elements gives the column a self-restoring property and makes it strong enough to cope with a variety of distortions (Symmons, 1979).

The same electromyographic studies have shown that as swimming speed increases deeper muscle layers are recruited which are not active during slow swimming. When swimming at higher speeds, a much greater volume of muscle is required to give the necessary force. The greatest part of the rest of the lateral muscle is the anaerobic fast white muscle, but five different intermediate categories of muscle between red and white have recently been identified in cod in layers immediately beneath the red muscle (Korneliussen, Dahl & Paulsen, 1978) and a similar situation exists in other species (Bone *et al.*, 1978).

The large volume of this white fast muscle is formed into complex-shaped myotomes, separated by myosepts. Each myosept is a sheet of collagenous fibres (seen as a boot and tongue shaped outline in Fig. 6). The muscle tissue of each myotome can be imagined filling the space between two adjacent myosepts with the muscle fibres approximately in line with the body's longitudinal axis.

It is notable how the red muscle in the surface layer has a simple segmental structure in contrast to the complex overlapping arrangement of the white muscle. A similar overlapped arrangement of the myotomes has been described in many fish species in particular by Greene & Greene (1913), Shann (1914) and Nursall (1956). The fibre direction, in the white part of the myotomes is more complex than in the superficial red muscle layer. The particular arrangement of the white muscle fibres, not quite in line with the fish's longitudinal axis, has been shown to give a similar degree of contraction and stretch during the swimming movements whatever their position in the myotome (Alexander, 1969; Kashin & Smoljaninov, 1969; Stelt, 1968). The number of fibres in any cross-section of a fish (two stippled areas Fig. 6) is not altered by the complicated attitudes of the myotomes. The muscle fibres bridge the gaps between two myosepts attaching firmly at each end, often at an acute angle to the plane of the myosept. These fibre-to-myosept angles become more acute in myotomes closer to the tail, partly as a result of the narrowing of the body while maintaining a similar degree of spacing and overlap of the myotomes.

The minimum part of the thicker white muscle layer that can shorten separately is one myotome. However with the sort of overlap of myotomes shown in Fig. 6 a complete transverse section of the cod is only shortened when more than four adjacent myotomes are in a contracted state at the same moment. Any transverse section of the body either side of this fully contracted zone will show partial shortening. When a wave of contraction moves along the body, involving complete contraction of each myotome (e.g. in high-speed swimming) there will be a progressive shortening and relaxation of each transverse section of the body due to this overlap (Willemse, 1966). To obtain the same smooth transition from relaxed to contracted state would involve a very complex arrangement of nerve–muscle control if the myotome overlaps did not exist (Willemse, 1966).

How does muscle contraction move the fish?

The simplest way of using the force from the contracting muscle is to pass it directly to the water. This could be achieved by pressing the bending part of the body wall and its local fins against the water as the wave of contraction moves backwards. This method is used by those ribbon-like fish such as eels. Typically the cylindrical body tapers only at the very end, bears continuous dorsal and ventral fins and has no distinct tail fin blade. The way in which the eel-like body reacts all along its length with the water is described by Gray (1933a).

In marked contrast to eels, it is claimed that some of the scombroid fish do shed nearly 100% of the propulsive force from the caudal fin (Fierstine & Walters, 1968). The same authors described the general adaptations of the anatomical structures particularly of the vertebral column. They show how the much overlapped myotomes of the opposing lateral muscles join into giant tendons which work the lunate, wing-like tail via highly developed tail hinge joints. The various members of the Scombridae, Istiophoridae and Xiphiidae which show the features described by Fierstine & Walters (1968) appear to have lost completely the use of waves in their swimming methods. Our records of mackerel swimming however show (Fig. 4) a body wave very similar to the cod and it seems there may be a continuous series of types from fish swimming by body wave propagation to those extreme examples discussed by Fierstine and Walters where the body wave system is abandoned and the myotomes are overlapped so much that they are worked as two opposing muscle units.

The swimming of fish like cod, salmon and mackerel represents an intermediate type between the eel-like and tuna-like extremes and the way in which muscle force is passed to the water shows complications related to this position. The tapered shape of the cod, salmon and mackerel puts the major bulk of the anaerobic fast swimming muscle well forward from the tail. Yet in this type of fish with a narrow tail stem and a large tail fin it is thought that a large but unknown fraction of the muscle energy, used to move the fish forwards during steady swimming, is shed into the water via the trailing edge of the tail fin. Measurements of swimming power made during steady swimming and based entirely on the movement of the tail blade, appear to give reasonable results (Lighthill, 1971; Wardle & Reid, 1977; Videler & Wardle, 1978). Following exhausting swimming, muscle samples from myotomes just behind the head, just ahead of the tail or from a middle part, all show depletion of glycogen and lactic acid increased to a similar level (Black, Robertson & Parker, 1961, table 3, p. 95, Kamloops trout). All this indicates that in these intermediate-type fish, a large part of the force required for swimming is developed in the thick anterior part of the body and passes along the moving body to the tail blade. How are forces transferred from the thick part to the tail tip?

All the muscle fibres on one side of the fish can contract simultaneously and bend the fish over its full length. In such a movement, which can be seen in a standing start, the muscle fibres in the head region take a direct part in moving the tail. During the progress of a wave of contraction (as shown in Fig. 2), zones of the body contracting more than one-half wavelength ($\frac{1}{2} \lambda_b$) anterior to the tail, are pulling in a direction opposite to the tail movement. Normally two zones and sometimes three zones are contracting on alternate sides of the body at the same time (see dotted lines in Fig. 2). Grillner & Kashin (1976) have detected electrical activity in such groups of myotomes contracting simultaneously. A group of myotomes contracting near the head will stretch those uncontracted myotomes nearer to the tail and there is no way in which this contraction will be transmitted by these relaxed and stretching muscle fibres to the tail blade. This argument is true until the wave of contraction reaches the last myotome where the contracting fibres pull directly on the tail blade fin ray bases (Videler, 1975 & 1977, Fig. 6).

The bulk of the force developed by the anterior myotomes is used to displace the mass of the body laterally at the contracting point. If the backward velocity of this wave of contraction is close to the forward speed, the body wall will not react to any large extent with the surrounding water, but the body mainly follows through a tunnel made by the head. During the passage of a wave the muscle forces cause a significant lateral movement of the body oscillating it through a mid-point. The myotomes contracting in sympathy will maintain and add to the amplitude of this lateral oscillation of the body mass. Gray (1933b) suggested that the mechanical stretching of the next myotome might excite this myotome to contract in an automatic way. There is however no evidence for such an automatic system generating the wave but the build-up of the wave by appropriate impulses from the nervous system would allow the variety of wave velocities observed. As the body gets narrower and the mass of each segment gets less, the amplitude of the movement grows.

The reacting surface of the last part of the body wall is small and convex but the tail fin with negligible mass displays a large vertical surface and makes the largest lateral displacements. The tail blade, moving rapidly through a wide amplitude, reacts with a large amount of water. In this way the energy derived from lateral swimming muscles in all parts of the body helps to generate a major propulsive force at the tail tip.

A number of authors have realised that the myosepts are formed from non-stretchable sheets of collagenous material to which the muscles are attached at an angle and their role as tendons have been discussed in a number of papers. Nursall (1956) pointed out the existence of lines of reinforcement in certain directions on the surface of the myosept and the inclusion in some cases of branching bones to strengthen the myosept sheets.

In our own observation of cod myotomes, the outer part of the myosept is firmly and directly attached to the skin in the median part of the lateral body wall (Fig. 5) (see also Videler, 1975, Fig. 19). The inner part of the myosept is folded abruptly towards the tail just before inserting on the axial skeleton. Shann (1914) described this region of the myotome referring to it as the anterior pyramid. Greene & Greene (1913) described a similar structure in the salmon. Nursall (1956, Figs. 8 & 9) shows this fold or anterior cone. It does not appear to have been realised previously, that this well-described fold of the myosept forms a simple hinge mechanism, freeing the myosept

Fig. 7. Schematic horizontal section through a fish like the cod shown in Fig. 6, bent by muscle contraction on one side. The section is made slightly above the median horizontal plane and shows how the forward pointing fold in each myosept, allows the myosepts to remain equal in length, be attached to the vertebral column and the skin, yet be non-stretchable during active bending.

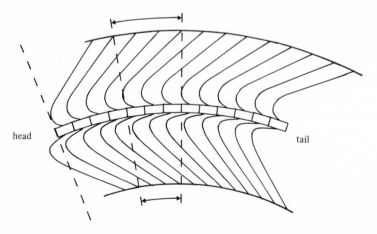

head   tail

from the effects caused by attachment along its inner edge. This freedom not only allows the fish to bend while retaining the non-stretching tendinous properties, but also frees the myosepts and allows direct transmission of forces from the contracting muscle fibres of the myotomes to the skin (Fig. 7). The criss-cross fibre structure of the fish skin has been described by Faure-Fremiet (1938), Bertin (1958) and Videler (1975). Videler (1975) showed that this fibre structure can be arranged either to transmit forces or to stretch according to the bulging of the muscles below. Wainwright, Vosburgh & Herbank (1978) have concluded that there is a similar function for the skin of elasmobranchs and have shown that pressure changes beneath the skin vary the skin transmission properties. Videler (1977, Fig. 6) described the joint between tail fin and body and showed that the skin and the last myotome are attached to the fin ray bases. Videler (1977, Fig. 7) shows how different forces pulling on the right and left side of the fin ray bases, cause the fin ray to bend. Tail fin rays will bend towards that side where the wave of contraction reaches the caudal peduncle. The last myotome and the skin by pulling on the fin ray bases on the same side will bend the rays in the direction of the motion and in doing so increase the effect of the fin on the water.

The physiological limits to swimming performance

Speed, power and muscle volume

The electromyographic measurements have shown that the red muscle layer is used during slow swimming and the much more massive white muscle is used only at higher speeds. At any steady swimming speed u, the power needed to overcome the drag is equal to $\frac{1}{2}\rho Su^3 \times 1.2 C_f$; where ρ is the water density, S the surface area of the fish and C_f the coefficient of friction: for further discussion see Wardle & Reid (1977). Some preliminary measurements of a cod swimming at different speeds showed that power used was more than cubed for each unit increase in speed (Wardle & Reid, 1977). Power is the energy used per second and is directly related to the volume of muscle contracting per second. The fish must therefore at least cube the volume of muscle used per second for each unit increase in velocity. The small volumes of slow muscle and intermediate muscle cannot help during the faster contractions and must be ignored when considering the total volume of muscle used. It is often pointed out that the fish is able to carry this neutrally buoyant mass of white muscle for only occasional use because its bulk helps to form the streamlined shape of the body.

Maximum speed

We have shown that waves of contraction of the lateral muscle passing backwards at a velocity v result in a forward swimming velocity u. The limit to the velocities v and therefore u is set by the time it takes for a piece of the muscle to shorten and not by any limit to the rate of passage of a wave of contraction along the myotomes. For example in the startle reaction all the segments on one side contract at nearly the same instant, probably synchronised by the Mauthner cells (Eaton, Bombardieri & Meyer, 1977; Zottoli, 1977). Isolated fish muscle fibres when stimulated by a single electrical impulse take a significant period to shorten. Fish 0.20 m long have white muscle contraction times as short as 0.02 s whereas the same muscle from fish 0.80 m long can take 0.05 s to shorten (Wardle, 1975).

The fastest efficient mode of swimming is clearly when the left myotome is just starting to relax as its opposing right myotome is just starting to contract. In this case one body wave or stride (λ_s in Fig. 2) would then be completed in a time equal to twice the shortening time of the fastest muscle. During steady swimming, when the speed of the wave backwards v is proportional to the resulting forward speed u, the distance moved forwards on completion of one stride is near $0.7 \times$ length (L) of the fish. By measuring the contraction time (Tm) of the isolated white swimming muscle the maximum swimming velocity u_{max} can be predicted where:

$$u_{max} = \frac{0.7 \times L}{2Tm}$$

The predicted velocities have been confirmed with television recordings of swimming fish (Wardle, 1975). Both length of fish and temperature affect the contraction time of the fast swimming muscle and the seasonal range of temperature encountered for example by a cod from 0 °C to at least 20 °C has a very large effect on the maximum swimming speed. For example cod ($L = 0.73$ m) should achieve a maximum swimming speed of 6.6 m s^{-1} at 20 °C, but only 4.3 m s^{-1} at 0 °C (Wardle, 1975). The deep white muscle temperature of fish is normally very close to the temperature of the water (Dean, 1976).

Endurance at maximum speeds

It has been repeatedly shown that the energy for maximum burst speed swimming is derived by the enzymes of the Embden–Meyerhof glycolytic pathway which convert glycogen stored in the white muscle cells to lactic acid (Black et al., 1961). These muscles can only work at full power when the glycogen store is in the rested condition. In trout two minutes of fast swimming can deplete the store by 50% (Black, Connor, Lam & Chiu, 1962). The maximum swimming speed is therefore available to the fish for quite short periods of minutes rather than hours and long rest periods of many hours are required in order to repeat bouts of fast swimming. As the muscle glycogen is depleted the fish is typically more reluctant to swim and makes greater efforts to seek shelter. When the muscle glycogen is exhausted the fish is unable to use the white fast muscle but must maintain a sheltered position, while the muscle regains its full performance potential (Black *et al.*, 1961). In the resting plaice, *Pleuronectes platessa*, 50–80% of the muscle glycogen was restored in eight hours (Wardle, 1978). Although the fastest bursts of swimming are of very short endurance they are of great survival value to the fish both in catching food and in escaping predators. Most fish live in situations where some form of shelter is within reach of a brief high-speed burst of swimming.

Cruising speeds

The red muscle just beneath the skin (see Figs. 5 & 6) is all the muscle required to swim at the cruising speed. The maximum oxygen intake of haddock (a fish like cod) is close to $300 \, \text{mg} \, O_2 \, \text{kg}^{-1} \, \text{h}^{-1}$ (Tytler, 1969) and of salmon $600 \, \text{mg} \, O_2 \, \text{kg}^{-1} \, \text{h}^{-1}$ (Brett & Glass, 1973) and more than $1000 \, \text{mg} \, O_2 \, \text{kg}^{-1} \, \text{h}^{-1}$ has been recorded in anaesthetised tuna (Stevens, 1972). These different ceilings limit the aerobic swimming performance. The limited haematocrit, heart capacity and gill area are all just matched and capable of carrying this oxygen to the aerobic muscle at the maximum rate. There is myoglobin in the muscle to bias the passage of the oxygen at the maximum rate to the muscle cells. In fish with higher sustained cruising rates like mackerel, all these aerobic features are set at a higher level in order to direct more oxygen from the water to the aerobic muscle. The power made available by this oxygen supply is applied by the red muscle to generate a slow body wave. The slower contraction properties of this red muscle are suited to making a slow body wave with a maximum wave speed related to this maximum cruising swimming speed. The balance between the maximum uptake of oxygen and the resulting aerobic swimming speed can change with both size and type of fish and temperature and was discussed in detail by Wardle (1977a).

Intermediate swimming speeds

In experimental conditions fish can be kept swimming for weeks on end at speeds below and up to their maximum cruising speed (Wardle,

1977a) but when made to swim at speeds only slightly above this established maximum cruising speed the endurance immediately drops to minutes. As the swimming speed increases the tail beat frequency rises smoothly maintaining a fixed relationship between the speed of the wave backwards and the speed of the fish forwards.

However when fish are kept swimming at these intermediate speeds, the swimming style appears quite soon to transfer from a steady fast tail beat to a burst-glide-burst-glide style while maintaining the original speed overall. This change in style of swimming is particularly well seen by divers observing fish maintaining station with the moving parts of towed fishing gears (Hemmings, 1973; Wardle, 1977b). It has also been repeatedly seen to develop in cod, salmon, saithe and haddock racing between feeding points in the experiments such as those described by Wardle & Kanwisher (1974) in the 10-m gantry tank at Aberdeen. This observation suggests that muscle fibres suited particularly to steady intermediate speed swimming are soon exhausted and the fish then resorts to using intermittent bursts of the fastest white muscle.

As mentioned above, the fish needs the largest proportion of its muscle volume in order to develop the required power for maximum swimming speed. The onset of the 'burst-glide' style of swimming suggests that the fish is unable to use the bulk of its white swimming muscle in a low-power steady-swimming mode, possibly because the muscle units are only arranged, by their innervation, in the large powerful blocks that are required for the fastest swimming. Use of smaller zones of white muscle in each myotome in a slower tetanic mode is not apparently possible. Bone et al. (1978) concluded from their electromyographic records that carp may be able to use their white muscle for slow steady swimming whereas Pacific herring and rainbow trout did not. This rather puzzling result shows that more experiments are required.

Increasing the distance moved for the same energy reserve

The maximum sustained cruising speed is the fastest speed that the fish can continue to use without tiring. Large fish (0.7–1.0 m) use a steady tail beat of just over 1 Hz and smaller fish (0.1–0.2 m) 3–4 Hz (Wardle, 1977a). Normally one might assume that this speed makes use of fuel from a relatively large store topped up by occasional feeding so that this fuel never runs out. Fish do migrate at times of poor feeding and can have fuel stores of up to 40% of the body muscle weight as fat, e.g. in herring (Blaxter & Holliday, 1963).

Individual fish have been tracked in their normal environments by adding ultra-sonic tags (for references see Tytler, Machin, Holliday & Priede, 1978). The typical voluntary day-to-day movements are much slower than maximum cruising and on average use a metabolic level at about 20% of the maximum aerobic level (Priede, 1977). The fish uses its maximum activity level during less than 10% of its life and it is argued that this is an important factor in reducing mortality and increasing survival of a species (Priede, 1977).

In a situation where the fuel is limited and there is no requirement to swim at the maximum cruising speed then it has been shown theoretically that several specific swimming strategies will increase, by up to five times, the distance covered using the same fuel. Weihs (1973a) has argued that a fish, using a steady tail beat at an energy level just twice the basal metabolic level (in his example this was about 25% maximum cruising speed), will swim twice the distance it would have covered if using the same amount of fuel at maximum cruising speed.

In another study Weihs (1974) showed that a fish swimming by alternating steady tail beats at maximum cruising speed frequency with periods of gliding with the body straight, could increase the distance covered by up to three times that moved at constant maximum cruising speed using the same fuel. Similar arguments were used to show that fish heavier than water may save energy by actively swimming upwards and then gliding downwards in a saw-tooth pathway (Weihs, 1973b). Fish swimming steadily in schools might by appropriate spacing in a diamond pattern make significant savings in energy for the distance moved (Weihs, 1973c). Weihs (1977) developed an optimum speed swimming model by examining the effect of size increase. Optimum speeds of fish 0.5 m, 1.0 m and 5 m long were 0.4 m s^{-1}, 0.5 m s^{-1} and 1.0 m s^{-1}. In practice the optimum speed appears to approach the maximum sustained cruising speed as size increases (Wardle, 1977a) but a great deal more data is required to establish both these estimates. These elegant models should clearly be re-examined as further exact data for drag, thrust and motor efficiency are acquired.

Estimating propulsive force

From high-speed cine- and videorecordings showing the dorsal view of a fish swimming past a camera in a fixed position, the speed and direction of movement of every part of the body can be determined. An example similar to that used to develop our basic model in Figs. 1 & 2 is given in Fig. 8. Successive positions of the head and the end of the tail blade of a cod are shown (Fig. 8a) and the instantaneous tail tip velocity is analysed (Fig. 8b). It is similarly possible to analyse velocities of every part of the body during the whole swimming cycle. Such knowledge of the movements of the body relative to the water can be used in different ways to estimate amounts of propulsive force.

Four approaches to analysing propulsive force

We will try to illustrate four approaches (a–d) by considering the tail blade end as representing an arbitrary piece of the surface of the fish. The movements of any piece of a surface in water can generate thrust in different ways:

(a) The tail blade piece encounters resistance when moved through the water. The resistance met by the component in tangential direction k is mainly caused by friction and is equal to kC_t where C_t is a coefficient of resistive force when the movement is tangential to the surface. The component in the negative X-direction (Fig. 8a) of this force is $kC_t \cos \theta$. Any movement of the tail blade through the water normal to the surface (vector w Fig. 8b) is resisted by force wC_n, where C_n is a coefficient of positive resistive force when movement is normal to the surface. The component in the positive direction of this force is $wC_n \sin \theta$. The instantaneous

Fig. 8. Two different ways to analyse the speed and direction of movement of the end of the tail blade of a fish. Further explanation is given in the text.

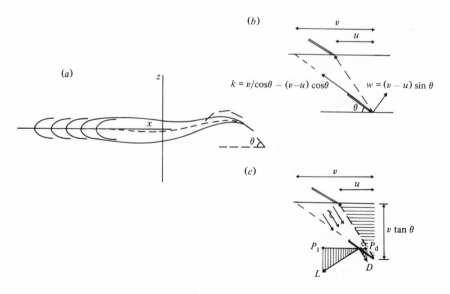

propulsive force along the X-axis resulting from the movement of the tail blade piece is

$$P = wC_n \sin \theta - k\, C_t \cos \theta$$

The values of C_n and C_t are difficult to obtain. Both coefficients are proportional to the density of the water, the surface area of the tail blade piece and the speed vectors w and k. The tangential coefficient of resistive force (C_t) depends particularly on the flow conditions in the boundary layer and on the Reynolds number of the tail blade piece. For further ideas on this resistive theory see Gray (1968) and Taylor (1952).

(b) A different approach is based on Lighthill's (1971) large-amplitude elongated-body theory for fish locomotion. This theory describes how a certain mass of water, the so-called virtual mass, is involved when the fish body moves. The inertia of this mass of water reacts against the movements of each part of the body surface. The sizes of these forces will be proportional to the rate of change of velocity of the body surfaces relative to the surrounding still water and to the virtual mass involved.

Here we will show how instantaneous propulsive force is thought to be generated by the end of the tail blade. Rate of change of momentum, generated by the wave-like movements of the whole body involve rate of change of velocity of each part of that body. This rate of change of momentum probably equals zero over a number of complete swimming cycles, because the velocity increases and decreases periodically in every part of the body and movements are supposed to be symmetrical about the X-axis.

Mean values of the instantaneous forces, generated by the end of the tail blade, could provide a good estimate of the mean thrust produced by the fish. Fig. 8b shows the two vectors k and w to be taken into consideration. The tangential movement with velocity k hardly evokes any reactive forces from the water. The viscous resistance between water and body is very small, especially when the surface is smooth. The virtual mass of water affected by k is very small and these reactive forces can be neglected. The situation is completely different in the case of component w (Fig. 8b). Here the virtual mass of water per unit length is

$$m = \pi\rho(s/2)^2$$

Where s is the height of the tail blade and ρ is the density of the water. The virtual mass of water in reaction with the tail blade per unit time is km (tangential velocity k is the distance moved in unit time) and the resulting total momentum in the w direction is kmw. Its component along the X-axis is:

$$kmw \sin \theta$$

In unit time this momentum imparts velocity to the mass of water involved, in a positive X-direction (Fig. 8a) and includes a propulsive force equal and opposite which gives forward velocity to the fish.

Lighthill (1970, equations 14 & 15) shows how (according to Bernoulli's equation) differences in velocity between fast water near the tail blade and motionless water at some distance, cause a pressure difference resulting in a pressure force equal to $\frac{1}{2}mw^2$ which subtracts from the thrust. This pressure difference occurs across a plane at right angles to the end of the tail blade, so the direction of the force $\frac{1}{2}mw^2$ is normal to this plane and tangential to the tail blade. The component of this force in the X-direction is $-\frac{1}{2}mw^2\cos\theta$. The instantaneous propulsive force from this approach is:

$$P = kmw\sin\theta - \tfrac{1}{2}mw^2\cos\theta$$

Development and application of this approach is included in Lighthill (1971), Wardle & Reid (1977) and Videler & Wardle (1978).

(c) In the third approach we treat the tail blade as a hydrofoil (Fig. 8c). The velocity and direction of the tail tip (the dashed vector Fig. 8c) is equal to the velocity at which the water passes the tail blade in the opposite direction, its magnitude is in terms of wave velocity v and forward velocity u:

$$\sqrt{(v^2\tan^2\theta + u^2)}$$

This water flow (three parallel arrows Fig. 8c) will generate a lift force in a direction perpendicular to the tail tip movement and a drag force in the direction of the flow. The lift force (L, Fig. 8c) is proportional to:

$$\tfrac{1}{2}\rho S_h(v^2\tan^2\theta + u^2)C_l$$

Where ρ is the density of the water, S_h is the surface area of the hydrofoil and C_l is the lift coefficient. The drag force (D, Fig. 8c) is proportional to:

$$\tfrac{1}{2}\rho S_h (v^2\tan^2\theta + u^2)C_d$$

C_d is the drag coefficient. The magnitude of the lift and drag coefficients depends on the shape of the profile of the fin, on the angle between the flow direction and the surface of the fin and on the conditions of the flow. The lift force makes a positive contribution (P_l, Fig. 8c) and the drag force a negative contribution (P_d, Fig. 8c) to the instantaneous propulsive force (P) along the X-axis. The hatched triangles in the diagram of Fig. 8c are similar which makes it possible to show that:

$$P = P_l - P_d = \tfrac{1}{2}\rho S_h \sqrt{(v^2\tan^2\theta + u^2)}(v\tan\theta\, C_l - uC_d)$$

Treatments of the hydrofoil or wing approaches to analysis of power output by fish have been made by Wu (1971), Chopra & Kambe (1977) and Lighthill (1977).

(*d*) These first three approaches consider only limited parts of the complex effect of the swimming movements of the fish on the fish itself and on the water. A more complete method requires exact knowledge of the movements of the water due to the movements of the fish. This knowledge could be gained by the use of flow visualisation techniques. Hydrodynamicists can provide an ultimate method to calculate propulsive force of a swimming fish as soon as the complete flow pattern along and behind the body of a fish, swimming steadily through otherwise undisturbed water, could be visualised. A true description of the flow gives knowledge of the pressure distribution around the fish and this added to the precise knowledge of the kinematics which we already have, could solve the main problem of fish locomotion. So far results from attempts to apply flow visualisation techniques have been disappointing.

We are grateful for a NATO grant No. 1127 assisting us with this cooperative work and for a Royal Society Fellowship in the European Science Exchange Programme to one of us (JJV). We value greatly discussions with Dr Saida Symmons and Regina Getreuer on points of fish anatomy and with Dr Aenea Reid on kinematic analysis.

References

Alexander, R. McN. (1969). The orientation of muscle fibres in the myomeres of fishes. *J. mar. biol. Assoc., UK,* **49**, 263–90.

Bainbridge, R. (1958). The speed of swimming of fish as related to size and to the frequency and amplitude of tail beat. *J. exp. Biol.* **35**, 109–33.

Bertin, L. (1958). Peau et pigmentation. In *Traité de Zoologie, Anatomy, Systematique, Biology*, vol. 13, part 1, ed. P. P. Grassé, pp. 433–58. Paris: Masson et Cie.

Black, E. C., Connor, A. R., Lam, K. & Chiu, W. (1962). Changes in glycogen, pyruvate and lactate in rainbow trout (*Salmo gairdneri*) during and following muscular activity. *J. Fish. Res. Board Can.* **19**, 409–36.

Black, E. C., Robertson, A. C. & Parker, R. R. (1961). Some aspects of carbohydrate metabolism in fish. In *Comparative Physiology of Carbohydrate Metabolism in Heterothermic Animals*, ed. A. W. Martin, pp. 89–122. Seattle: University of Washington Press.

Blaxter, J. H. S. & Holliday, F. G. T. (1963). The behaviour and physiology of herring and other clupeids. In *Advances in Marine*

Biology, vol. 1, ed. F. S. Russell, pp. 262–372. London and New York: Academic Press.

Blight, A. R. (1977). The muscular control of vertebrate swimming movements. *Biol. Rev.* **52**, 181–218.

Bone, Q., Kiceniuk, J. & Jones, D. R. (1978). On the role of the different fibre types in fish myotomes at intermediate swimming speeds. *Fish. Bull., US,* **76**, 691–9.

Brett, J. R. & Glass, N. R. (1973). Metabolic rates and critical swimming speeds of Sockeye salmon (*Oncorhynchus nerka*) in relation to size and temperature. *J. Fish Res. Board Can.* **30**, 379–87.

Chopra, M. G. & Kambe, T. (1977). Hydromechanics of lunate-tail swimming propulsion part 2. *J. Fluid Mech.* **79**, 49–71.

Dean, J. M. (1976). Temperature of tissues in freshwater fishes. *Trans. Am. Fish. Soc.* (1976), 709–11.

Eaton, R. C., Bombardieri, R. A. & Meyer, D. L. (1977). The Mauthner-initiated startle response in teleost fish. *J. exp. Biol.* **66**, 65–81.

Faure-Fremiet, E. (1938). Structure du derme téliforme chez les Scombridés. *Arch. Anat, Microsc., Paris,* **34**, 219–30.

Fierstine, H. L. & Walters, V. (1968). Studies in locomotion and anatomy of scombroid fishes. *Mem. south. California Acad. Sci.* **6**, 1–31.

Gray, J. (1933a). Studies in animal locomotion. I. The movement of fish with special reference to the eel. *J. exp. Biol.* **10**, 88–104.

Gray, J. (1933b). Studies in animal locomotion III. The Propulsive mechanism of the whiting (*Gadus merlangus*). *J. exp. Biol.* **10**, 391–400.

Gray, J. (1968). *Animal Locomotion*. London: Weidenfeld & Nicolson.

Greene, C. W. & Greene, C. H. (1913). The skeletal musculature of the king salmon. *Bull. US Bur. Fish.* **33**, 21–60.

Grillner, S. & Kashin, S. (1976). On the generation and performance of swimming in fish. In *Neural Control of Locomotion*, ed. R. M. Herman, S. Grillner, P. S. G. Steen & D. G. Stuart, pp. 181–201. New York and London: Plenum Press.

Hemmings, C. C. (1973). Direct observation of the behaviour of fish in relation to fishing gear. *Helgoländer wiss. Meeresunters.* **24**, 348–60.

Kashin, S. M. & Smoljaninov, V. V. (1969). On the question of the geometrical arrangement of body muscles in fish. *Vop. Ikhtiol.* **9**, 1139–42. (In Russian.)

Korneliussen, H., Dahl, H. A. & Paulsen, J. E. (1978). Histochemical definition of muscle fibre types in the trunk musculature of a teleost fish (Cod, *Gadus morhua, L.*). *Histochemistry*, **55**, 1–16.

Lighthill, M. J. (1970). Aquatic animal propulsion of high hydromechanical efficiency. *J. Fluid Mech.* **44**, 265–301.

Lighthill, M. J. (1971). Large amplitude elongated-body theory of fish locomotion, *Proc. R. Soc., Ser., B,* **179**, 125–38.

Lighthill, M. J. (1977). Mathematical theories of fish swimming. In *Fisheries Mathematics*, ed. J. H. Steele, pp. 131–44. London, New York and San Francisco: Academic Press.

Nursall, J. R. (1956). The lateral musculature and the swimming of fish. *Proc. zool. Soc., Lond.* **126**, 127–43.

Pitcher, T. J., Partridge, B. L. & Wardle, C. S. (1976). A blind fish can school. *Science, NY,* **194**, 963–5.

Priede, I. G. (1977). Natural selection for energetic efficiency and the relationship between activity level and mortality. *Nature, Lond.* **267**, 610–11.

Rayner, M. D. & Keenan, M. J. (1967). Role of red and white muscles in the swimming of the skipjack tuna. *Nature, Lond.* **214**, 392–3.

Shann, E. W. (1914). On the nature of the lateral muscle in Teleostei. *Proc. zool. Soc., Lond.* (1914), 319–37.

Stelt, A. Van der, (1968). Spiermechanica en Myotoombouw bij Vissen. Thesis, Amsterdam. (Summary in English.)

Stevens, E. D. (1972). Some aspects of gas exchange in tuna. *J. exp. Biol.* **56**, 809–23.

Symmons, S. (1979). Notochordal and elastic components of the axial skeleton of fish and their function in locomotion. *J. Zool., Lond.* **189**, 157–206.

Taylor, G. (1952). Analysis of the swimming of long narrow animals. *Proc. Ry. Soc., Ser. A,* **214**, 158–83.

Tytler, P. (1969). Relationship between oxygen consumption and swimming speed in the haddock, *Melanogrammus aeglefinus. Nature, Lond.* **221**, 274–5.

Tytler, P., Machin, D., Holliday, F. G. T. & Priede, I. G. (1978). A comparison of the patterns of movement between indigenous and displaced brown trout (*Salmo trutta L*) in a small shallow loch. *Proc. R. Soc. Edinb. Ser. B,* **76**, 245–68.

Videler, J. J. (1975). On the interrelationships between morphology and movement in the tail of the cichlid fish *Tilapia nilotica* (L.). *Neth. J. Zool.* **25**, 143–94.

Videler, J. J. (1977). Mechanical properties of fish tail joints. *Fortschr. Zool.* **24**, 183–94.

Videler, J. J. & Wardle, C. S. (1978). New kinematic data from high speed cine film recordings of swimming cod (*Gadus morhua*) *Neth. J. Zool.* **28**, 465–84.

Wainwright, S. A., Vorsburgh, F. & Hebrank, J. H. (1978). Shark skin: function in locomotion. *Science, NY,* **202**, 747–9.

Wardle, C. S. (1975). Limit of fish swimming speed. *Nature, Lond.* **225**, 725–7.

Wardle, C. S. (1977a). Effects of size on swimming speeds of fish. In *Scale Effects in Animal Locomotion*, ed. T. J. Pedley, pp. 299–313. New York and London: Academic Press.

Wardle, C. S. (1977b). *Divers Observe Seine Nets Fishing*. 16 mm film, 30 min. Available: DAFS Marine Laboratory, PO Box 101, Aberdeen, UK.

Wardle, C. S. (1978). Non-release of lactic acid from anaerobic swimming muscle of plaice *Pleuronectes platessa L.*: A stress reaction *J. exp. Biol.* **77**, 141–55.

Wardle, C. S. & Kanwisher, J. W. (1974). The significance of heart rate in free swimming cod, *Gadus morhua*: Some observations with ultra-sonic tags. *Mar. Behav. Physiol.* **2,** 311–24.

Wardle, C. S. & Reid, A. (1977). The application of large amplitude elongated body theory to measure swimming power in fish. In *Fisheries Mathematics*, ed. J. H. Steele, pp. 171–91. London, New York and San Francisco: Academic Press.

Weihs, D. (1973a). Hydromechanics of fish schooling. *Nature, Lond.* **241,** 290–1.

Weihs, D. (1973b). Optimal cruising speed for migrating fish. *Nature, Lond.* **245,** 48–50.

Weihs, D. (1973c). Mechanically efficient swimming techniques for fish with negative buoyancy. *J. mar. Res.* **31,** 194–209.

Weihs, D. (1974). Energetic advantages of burst swimming of fish. *J. theor. Biol.* **48,** 215–29.

Weihs, D. (1977). Effects of size on sustained swimming speeds of aquatic organisms. In *Scale Effects in Animal Locomotion*, ed. T. J. Pedley, pp. 299–313. New York and London: Academic Press.

Willemse, J. J. (1966). Functional anatomy of the myosepta in fishes. *Proc. k. ned. Akad wet.* **69,** 58–63.

Wu, T. Y. (1971). Hydromechanics of swimming fishes and cetaceans. *Adv. appl. Math.* **11,** 1–62.

Zottoli, S. J. (1977). Correlation of the startle reflex and Mauthner cell auditory responses in unrestrained goldfish. *J. exp. Biol.* **66,** 243–54.

H. C. BENNET-CLARK

Aerodynamics of insect jumping

Introduction

When a body moves through a fluid, its motion is impeded by a drag force whose magnitude is dependent on properties of the fluid. At relatively high velocities or with large bodies, the drag on the body can be reduced by streamlining, but this is only effective where the major component of the drag force acts against the inertia of the medium. At lower velocities or with smaller bodies viscous effects become increasingly important and the shape of the body becomes less important as a means of minimising drag.

Since the moving body encounters a drag force that increases as the square of its velocity, drag acts as an effective limit to the maximum speed that the body can attain because the power available will be proportional to the mass of the body or the cube of length but the drag force, which acts on the frontal area of the body, is proportional to the square of length. Small bodies thus encounter relatively higher drag forces than larger ones and so will tend to achieve lower limiting velocities than larger ones; this is observed in animals as diverse as birds (Tucker, 1977) and water-beetles (Nachtigall, 1977).

Jumping animals encounter the effects of drag in a rather different way to flying or swimming animals because the energy used in the jump is produced during the initial impulse rather than continuously. As a result, the velocity and hence drag forces are greatest just after the impulse and fall to zero at the top of a vertical jump. Jumping animals may be unable to control their attitude during the trajectory, unlike flying or swimming animals, and so may be unable to exploit streamlining as a means of reducing drag.

In this review, I consider the effects of drag on the jumping performance of animals of different sizes, the effect of body shape on the drag and how scale effects may effect the optimal design strategy for reducing aerodynamic drag. Some of the conclusions have implications beyond the special case of the jumping animal.

The effect of air resistance on jumping animals

Theoretical

In the practical case, a body may maintain a constant attitude or may rotate or oscillate in flight. As a result, the effective frontal area, A, which the body presents normal to the direction of flight and the drag coefficient of the body, C_D, which depends on the body's shape, may vary during flight.

To indicate the relative magnitude of the effects, it is more convenient to regard both A and C_D as constant, even though in a practical jumping animal, both of these terms may vary. For theoretical reasons, too, it is more convenient to confine an examination of the effects of air resistance to the special case of a vertical jump, for which the equations of motion can be solved easily.

A jumping animal takes off at a velocity, v, determined by the energy, E, that it can convert into kinetic energy during the impulse as follows:

$$v = \sqrt{\frac{2E}{m}} \tag{1}$$

where m is the mass of the body. The height in a vertical jump, h_v, that is attained *in vacuo* is determined by the retarding acceleration of gravity, g_n, on the body thus:

$$h_v = \frac{E}{m\,g_n} = \frac{v^2}{2\,g_n} \tag{2}$$

Given an energy: mass ratio of 20 J kg^{-1} which is a probable upper limit, this height will not exceed 2 m (Bennet-Clark, 1977) and is usually less.

When drag forces are taken into account, it is found that they act in the same sense as gravity to further restrict the height attained in the upward phase of the jump but act against gravity, restricting the velocity in the falling phase of the jump.

In the upward moving phase, the equation of motion of a body moving vertically in a fluid of density ρ is:

$$m\,\frac{d^2x}{dt^2} + \frac{C_D\rho A}{2}\left(\frac{dx}{dt}\right)^2 + mg_n = 0 \tag{3}$$

where x is the distance above the ground and t is the time since leaving the ground.

Rearrangement of equation 3 for the case of constant C_D, A and ρ gives the height, h_a, attained by the body in air.

$$h_a = \frac{m}{C_D \, \rho \, A} \, \ln \left(\frac{C_D \rho \, A \, v^2}{2 \, m \, g_n} + 1 \right) \tag{4}$$

We may combine equations 2 and 3 to give:

$$\frac{h_a}{h_v} = \frac{m}{C_D \, \rho \, h_v \, A} \times \ln \left(\frac{C_D \, \rho \, h_v \, A}{m} + 1 \right) \tag{5}$$

Equation 5 is plotted in Fig. 1 for several values of C_D and shows that the problem can be expressed in terms of three dimensional parameters; h_a/h_v (which is defined as the jump efficiency), the group $\rho h_v A/m$, and C_D (which is a function of the body shape). The values of the first two parameters can be obtained experimentally and, by reference to Fig. 1, the value of C_D may be determined.

In the downward moving phase, the body accelerates until the drag force equals the body's mass:

$$\frac{C_D \rho A}{2} \left(\frac{dx}{dt} \right) = m g_n \tag{6}$$

which may be arranged to give dx/dt or velocity, v, for the terminal condition of a body falling in air:

$$v = \sqrt{\frac{2 \, m \, g_n}{C_D \, \rho \, A}} \tag{7}$$

Fig. 1. The general case of text equation 5, showing how jump efficiency, defined as height attained in air/height *in vacuo* (h_a/h_v) varies with the group $\rho h_v A/m$ for various values of the drag coefficient, C_D (redrawn from Bennet-Clark & Alder (1979).

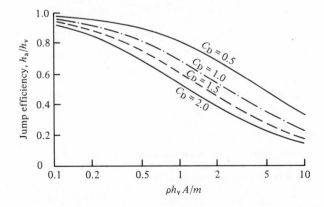

As with the upward moving phase, the performance is limited by the frontal area to mass ratio of the body and by its drag coefficient, C_D.

Aerodynamic drag of insects' bodies

In a series of experiments described more fully elsewhere (Bennet-Clark & Alder, 1979), the bodies of insects were propelled vertically by means of a spring gun which could be adjusted to propel bodies at various values of initial velocity equivalent to specific values of height *in vacuo*, h_v. This was used to simulate jumps of different energy to mass ratios (equation 2) by insects of different sizes and hence of different frontal area to mass ratios. Frontal area in square millimetres viewed normal to the horizontal longitudinal plane is approximately $3.8 \times$ mass in $mg^{0.67}$. Fig. 2 shows how jump efficiency (h_a/h_v) falls with decreasing insect size; some dimensions of the insects used are shown in Table 1. It should be noted that differences in the drag coefficient recorded in Table 1 are subject to errors in

Fig. 2. Plots of jump efficiency, h_a/h_v, against frontal area to mass ratio for the bodies of the fleas *Ctenophthalmus* and *Hystricopsylla*, *Drosophila* without wings and larvae of *Schistocerca*. The bodies were projected upwards at velocities equivalent to various values of height *in vacuo*, h_v, and the height attained in air, h_a, was observed. The lines show the value of h_a/h_v calculated for bodies with a drag coefficient, C_D, of unity projected at various values of h_v. Redrawn from Bennet-Clark & Alder (1979.)

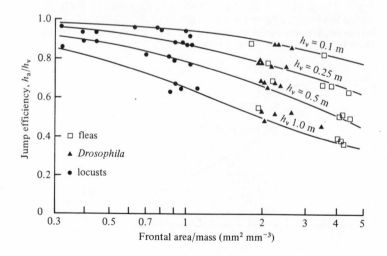

the estimation of the frontal area of the insects' bodies and so precise comparisons cannot be made between species.

It is clear from these experiments that aerodynamic drag is an important factor and that while a large locust may achieve a jump efficiency of over 0.8 when jumping at a velocity that would produce a height *in vacuo* of 1 m, a flea only achieves an efficiency of 0.4 – or travels half as far – for the same specific energy expenditure. Fleas, of course, achieve far higher efficiencies if they set themselves lower targets; a flea weighing about 0.5 mg can jump at the same efficiency of 0.8 with a value of h_v of 0.1 m as a locust weighing 0.5 g making a jump h_v of 1 m (see Figs. 1 and 2). Since the body of such a flea is about 2 mm long, it moves about fifty times its own length, a comparable proportional range to that of a 30 mm long 0.5 g locust. It is only when absolute performance limits are compared that the jump of a tiny insect appears ineffective and inefficient.

This apparent inefficiency is highlighted when estimates are made of the performance of various animals jumping with a specific energy of 20 J kg^{-1} (Table 2) which is the probable upper performance limit for a single muscle-powered impulse.

Drag is clearly of little practical importance either to larger jumping insects or to jumping mammals but its effects could act as an important evolutionary pressure in smaller jumping animals. Here body plans offering lower drag may be expected.

In other experiments, using *Drosophila melanogaster*, 1st instar locusts *Schistocerca gregaria* and the blow fly *Calliphora erythrocephala*, the effects of wings and legs on the aerodynamic drag were examined. The results of these experiments (Table 3) show that wings have a large effect on drag but

Table 1. *Some dimensions of insects' bodies and their drag coefficients*

Insect	Typical mass (mg)	Frontal area per unit mass ($mm^2 \text{ mg}^{-1}$)	Mean drag coefficient, C_D
Fleas:			
Ctenophthalmus	0.32	3.5	0.96
Hystricopsylla	1.9	1.8	1.00
Flies (wings removed):			
Drosophila	1.0	2.3	1.18
Calliphora	55	0.9	0.95
Locusts:			
Schistocerca 1st instar	15	1.0	1.08
4th instar	440	0.32	0.88

Table 2. *Jumping performance of animals of different sizes jumping at an initial energy of 20 J kg^{-1}, giving a height in vacuo, h_v, of 2 m*

Animal	Mass	Frontal area per unit mass (mm² mg^{-1})	Height in air h_a (m)	Jump efficiency h_a/h_v	Reynolds number, Re
Flea, *Ctenophthalmus*	0.3 mg	3.5	0.53	0.27	290
Locust, *Schistocerca*					
1st instar	15 mg	1.0	1.02	0.51	1200
4th instar	0.44 g	0.32	1.48	0.74	3000
Bush baby, *Galago senegalensis*	200 g (est.)	0.025 (est.)	1.93	0.97	20000

It is assumed that the drag coefficient, C_D, is 1 in all cases.

Table 3. *Effect of wing and leg amputation on weight and mean drag coefficient, C_D*

Insect	C_D intact	C_D without wings but with legs	Weight of wings as percentage of body weight	C_D, without wings, and without legs	Weight of legs as percentage of original body weight
Drosophila adults	1.58 (s.d. 0.35)	1.18 (s.d. 0.24)	2.3%	0.89 (s.d. 0.12)	7.2%
Calliphora adults	1.34 (s.d. 0.52)	0.95 (s.d. 0.10)	3.6%	0.84 (s.d. 0.08)	5.1%
Schistocerca 1st instar	1.08 (s.d. 0.21)	—	—	0.73 (s.d. 0.19)	25%

are light in weight; as wings are designed to have a large surface area to mass ratio, this is not surprising but, in the special case of jumping, such large light unused structures are clearly disadvantageous. The amputation of legs tends to be less effective as a means of reducing drag; in locusts, in particular, the hind legs are large and heavy because they contain the muscles used to provide the energy for jumping and so they contribute both to the mass of the animal and, to a small extent, to its frontal area. As a result, the lightened legless bodies of locusts tend to attain similar heights to the intact insects when propelled at the same initial velocity.

An extension of these findings is that winged insects will tend to fold their wings tightly both if they jump and if they are exposed to windy conditions where it may be disadvantageous to present large uncontrollable surfaces. The extreme solution is to become apterous.

Aptery is seen in a wide range of jumping insects, particularly among forms that are small or that are highly specialised for jumping. While Table 4 is not comprehensive it shows that aptery occurs among many small jumping insects and in those that are highly specialised for jumping.

Table 4. *Adaptations of some jumping insects*

Order	Family or genus	Length	Adaptations	Comments
Collembola	*Sminthurus*, etc.	3 mm	apterous globular	primitive aptery, small
Orthoptera	(larvae)	2–3 mm upwards	apterous	small, jump well
	Chorthippus	25 mm	wings reduced or vestigial	jump well
	Tetrix	10 mm	apterous	jump well
Homoptera	Cercopidae	up to 15 mm	wings fold tightly	jump well
	Jassidae	up to 8 mm	wings do not fold tightly	jump then fly
Mecoptera	*Boreus*	3 mm	wings vestigial	small
Diptera	many groups	2–20 mm	retain wings, jump at start of flight	jump only a few millimetres
Siphonaptera	most groups	mostly under 3 mm	apterous, globular	small, jump well
Coleoptera	*Halticus*	to 4 mm	wings under elytra, globular	small, jump well

Aerodynamic drag of bodies of different shapes and densities

Since jumping insects may be small and tend to move at low velocities, they tend to operate at low Reynolds numbers. In this regime, it is not possible to attain the low drag coefficients that can be attained by streamlined bodies at Reynolds numbers of 10^6; fleas jumping to heights of around 0.1 m have Reynolds numbers (Re) of about 100 while locusts jumping to heights of around 0.5 m have Re round 1500. At low Reynolds numbers, bodies tend to orient themselves so that the drag is maximal (Becker, 1959) and so the minimum drag is attained by bodies with the lowest frontal area to mass ratios.

This has been examined experimentally using models of different shapes and different frontal area to mass ratios (Bennet-Clark & Alder, 1979). In one series of experiments, models were made from balsa wood, of density around 0.1 mg mm^{-3}, and of pine wood of density around 0.5 mg mm^{-3}. The models were made in two sizes and either made as close approximations to spheres or as cylindrical rods of length to diameter ratio of about five to one. As these were cut by scalpel from the wood, it was not possible to obtain really accurate shapes but the results none the less were suggestive. In all cases, the drag coefficient of the spherical models was less than that of the long cylindrical models of similar weight, either around 0.5 mg or 4 mg. These small, rather rough models operated at similar Reynolds numbers to those of jumping fleas (Table 5).

In another series of experiments, small gelatine medicine capsules were modified so as to make a set of cylinders of different lengths but with hemispherical ends and constant diameter. Because the capsules were accurately formed, their frontal areas could be measured with a high degree of precision and because the two halves of the capsules could be separated, they could be cut with scissors and then reassembled with internal loads of different masses. Three lengths of capsule were used and they were loaded to give models of similar density to the bodies of insects. The results of experiments in which the capsules were projected upwards at a value of h_v of 1.04 m are shown in Table 6. In all cases the initial Reynolds number was 1800.

Making the capsules longer for a given mass has two effects, both of which are relevant. Increasing the length increases the drag coefficient directly. Increasing the length also increases the frontal area to mass ratio. The overall effect is that the shortest model, at a weight of 104 mg attains a height in air, h_a, of 0.88 m and an efficiency, h_a/h_v, of 0.85, while the medium-length model at a weight of 105 mg attains h_a of 0.74 and efficiency of 0.71 and the

Table 5. *Size and aerodynamic performance of wooden models*

Type of model	Mass (mg)	Ratio of maximum and minimum dimensions	Frontal area per unit mass (mm² mg⁻¹)	Jump efficiency, $h_v = 1\,m$	Mean drag coefficient, C_D	Max. Reynolds number, Re
Balsa ball	4.1	1.03:1	3.16	0.5	0.62 (s.d. 0.08)	1200
Balsa ball	0.55	1:1	5.15	0.38	0.66 (s.d. 0.08)	540
Pine ball	4.2	1.27:1	1.07	0.72	0.67 (s.d. 0.04)	660
Pine ball	0.69	1.33:1	2.19	0.56	0.70 (s.d. 0.03)	360
Balsa rod	4.2	6.9:1	5.30	0.35	0.80 (s.d. 0.05)	570
Balsa rod	0.46	5.4:1	11.7	0.23	0.71 (s.d. 0.06)	300
Pine rod	4.26	6.24:1	1.80	0.57	0.87 (s.d. 0.03)	330
Pine rod	0.57	4.41:1	3.58	0.43	0.85 (s.d. 0.03)	200

longest model, at the similar but rather lower weight of 98 mg only attains h_a of 0.6 m and an efficiency of 0.58.

The amputation experiments with insects and the two series of experiments with models suggest that there are a number of complementary strategies that can be adopted to reduce drag. The insect can reduce the number and size of protuberances from the body. The insect can also reduce its length until it approximates to a sphere, and it can reduce its frontal area to mass ratio for a given mass both by increasing its body density and by adopting a spherical shape.

Many larger flying insects have extensive air sacs, which tend to reduce overall body density but to allow external streamlining without weight penalty. This could be disadvantageous in a jumping insect which was unable to control its attitude to take advantage of such streamlining or which operated at such low Reynolds numbers that streamlining was relatively ineffective; in equation 5, frontal area, A, and drag coefficient, C_D, act in opposition so any improvement in efficiency brought about by reduction of C_D might be more than offset by the inevitable resultant increase of frontal area in a small jumping animal.

It is thus not surprising that many small jumping insects have rather short and globular bodies (Table 3) and that the tracheal systems of some of the smaller jumping insects should not have the extensive air sacs so typical of larger flying insects; this is true of fleas (Wigglesworth, 1935) Collembola

Table 6. *Aerodynamic performance of gelatine capsules*

Shape: length and diameter	Frontal area (mm²)	Mass (mg)	Frontal area per unit mass (mm² mg⁻¹)	h_a, in metres, when h_v is 1.04 m	Mean C_D all masses
Short		35	1.5	0.7	
9.7 × 6.2 mm	52	52	1.0	0.76	0.60
		104	0.5	0.88	(s.d. 0.05, n = 32)
		260	0.2	0.97	
Medium		59	1.8	0.63	
18.2 × 6.2 mm	105	105	1.0	0.74	0.73
		210	0.5	0.87	(s.d. 0.07, n = 32)
		525	0.2	0.95	
Long	182	98	1.86	0.60	0.79
30.5 × 6.2 mm		182	1.0	0.73	(s.d. 0.05, n = 32)
		364	0.5	0.84	
		910	0.2	0.94	

(Davies, 1927) and larval locusts (Clarke, 1957; H. C. Bennet-Clark, unpublished).

Reduction of drag

Optimal body shape as a function of size

At high speed or with large animals Reynolds numbers as high as 10^6 occur commonly. At these high values of Re, turbulence can occur in the flow, dramatically increasing drag so that streamlining, elimination of surface roughness and travel in a constant orientation all become important in minimising drag. Below values of Re of about 1000, the drag coefficient of streamlined bodies of all shapes tends to rise (Fig. 3) because viscous forces

Fig. 3. Plots of drag coefficient, C_D, against Reynolds number, Re, for bodies of various shapes. The short cylinder has a diameter to length ratio of 1:5. Data from various sources; for spheres, discs, streamlined bodies and short cylinders, redrawn from Leyton (1975), for 2:1 ellipsoids redrawn from Nachtigall (1977) and for *Drosophila* bodies calculated and drawn from data of Vogel (1965).

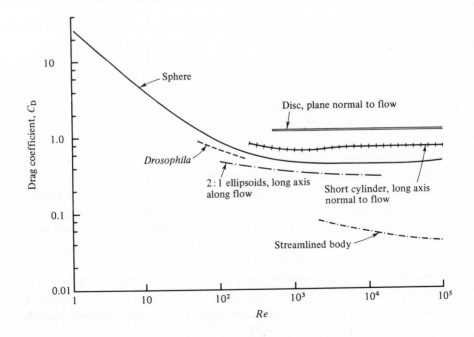

become increasingly important. At these low Reynolds numbers, the flow tends to be entirely laminar and the drag results from skin friction; drag then tends to be proportional to the gross diameter so a streamlined shape becomes less important. Surface irregularities become effectively obliterated by the relatively thick boundary layer in small animals.

Animals operating at higher Reynolds numbers tend to be well streamlined, but turbulence tends to occur and designs to minimise its effects are common (see e.g. Alexander, 1971). At rather lower values of Re, between 10^3 and 10^5, as in many birds and the larger insects, streamlining is clearly advantageous and turbulence may be avoided. Thus birds (Re above 10^4), Sphingid moths, dragonflies and locusts (Re between 2000 and 5000) all have bodies that are more or less streamlined without obvious protuberances, fusiform shape and a more or less pointed tail. At rather lower values of Re, the body shape shown by flies and bees is associated with the fastest flight. In this regime, Re 500–2000, the body tends to be relatively short and blunt, with a rounded tail. Because the boundary layer of body hairs can be relatively large, head, thorax and abdomen tend to be separated one from another by distinct necks and the space is smoothed over by specific arrangements of bristles pointing backwards over the space and presumably trapping the air in the space; such an arrangement will not appear rough unless the distance between bristles and their obtrusion into the air flow exceeds a critical value, d, given by:

$$d = \frac{58\,\mu}{\rho\,v} \tag{6}$$

where v is the velocity, ρ is the density and μ is the viscosity of the medium (Leyton, 1975). At a velocity of 4 m s^{-1}, attained by some flies and bees, the distance is 0.22 mm, while at 1 m s^{-1}, attained by the smaller flies, it rises to 0.87 mm. It is clear that the insect solution to the problem of aerodynamic drag will depend on size and Reynolds number.

The largest insects have smooth contours and a fineness ratio (the ratio between the length and the maximum width) of from 4:1 (Sphingid moths) and 8:1 (Anisopteran dragonflies) which is in the same range as that of bodies of fish and of many birds (see, e.g. Alexander, 1971). These large insects also tend to have dense but fine and flexible cuticular bristles, doubtless a convergent adaptation to the fine surface feathers on the bodies of birds. Both birds and large flying insects have extensive air sacs which could act as compensation spaces to accommodate changes in the dimensions of internal organs without change of the external contours.

Rather smaller and slower flying insects tend to have fineness ratios of between 4:1 for honey bees and wasps and 2.5:1 for some flies and bumble bees. With the smaller flies, the bristles tend to become sparser, as their effective boundary layer becomes proportionately bigger. Flies of the size of *Musca*, the house fly, still have extensive air sacs in both thorax and abdomen so it appears that the external form may still be important, though possibly less important than in larger insects.

Among the very smallest flying insects, bristles may or may not be present. It seems likely that even the necks between the different tagmata will not be large enough to cause separation of the air flow and that bristles might merely have the effect of increasing the thickness of an already expensive boundary layer. Here the evolutionary pressure will be to minimise both the frontal area and the surface area of the body because of the increasing frictional drag due to viscous effects. The smallest flying insects do not have extensive abdominal air sacs (e.g. Miller, 1950) even though there may be thoracic air sacs, presumably for respiratory reasons.

In the tiniest insects, the body may appear superficially to be bizarrely unstreamlined. Chalcid and Cynipid wasps, for example, often have humped bodies and spherical tagmata as do those of some of the smaller dipteran flies while the bodies of aphids are nearly spherical, with fineness ratios of about 1.5:1. In such insects, the Reynolds number of the body may be as low as 20–50 and the compromise between reduction of the form drag by streamlining and reduction of viscous drag by reduction of surface area becomes weighted towards reduction of the surface area by becoming spherical. Further, the effect of surface protuberances is merely to increase the thickness of the viscous boundary sub-layer and to effectively increase the frontal area and drag; such structures therefore tend to be lost.

Unfortunately, there is a lack of experimental work on the fluid dynamics of small bodies at Reynolds numbers in the range from 1 to 1000. The most detailed are those of Nachtigall (1977) for water-beetles. The largest, some 35 mm long, move at up to 0.4 m s^{-1} at Reynolds numbers of up to $14\,000$; for these he calculates drag coefficients of around 0.3. These parameters change for the smallest Dytiscids, only 2 mm long which achieve 45 mm s^{-1} at *Re* about 90, with a C_D of 0.8. In the larger insects, in water, the depth of the laminar sub-layer is about 0.15 mm, though it increases to over 1 mm for the smallest beetles studied. It is therefore not surprising that the largest water beetles have very smooth surfaces and surface hairs are largely lost (W. Nachtigall, personal communication) though their bodies appear to be rather broader than those of flying insects or fishes operating at similar values of *Re*. The lack of surface hairs in smaller water-beetles may merely reflect the trend to reduce frontal area that appears in the smallest flying insects.

Working with the bodies of *Drosophila virilis*, Vogel (1965) found that the drag was about 0.8 of that of a sphere of equivalent volume. Experiments in which he removed parts of the body tended to reduce mass and, to a lesser extent drag, but effectively to increase the drag coefficient. He concluded that the body form was closely optimised to minimise drag over the range of Reynolds numbers over which the animal operates, Re 60–300. Most significantly he noted that the drag of the bodies rose markedly when the direction of flow was reversed, suggesting that the body is optimised, not only to minimise drag but for a particular direction and attitude of travel.

Contrasts between jumping and flying animals

Throughout the preceding section, I have assumed that the animal can control the attitude of its body so as to orient its axis of minimum drag along the axis of its movement. This may not be possible in an animal moving ballistically – as in jumping. In such situations, the design of the body, as a projectile, may differ from that of the body as a fuselage.

The larger jumping insects, such as locusts, tend to have fairly long bodies – the fineness ratio for a locust is about 5.5:1 for late larvae and adults, falling to 4:1 for the smallest larvae. At the start of a jump, the Reynolds number is between 1500 for an adult and 300 for a first instar larva. At such values of Re, both change of shape and attitude are likely to lead to dramatic change of C_D. Jumping attitude in adult locusts appears to be partially controlled by a combination of mechanisms: the centre of gravity of the body is close to the centre of rotation of the hind coxae so the point of application of the jump impulse produces little pitching movement on the body; the body tapers from a relatively broad, rounded head, is widest around the metathorax – and centre of gravity – and has a long abdomen which tapers to about half the diameter of the head. The centre of pressure of the body is probably behind the centre of gravity and so the body may align itself along the direction of travel – or at least not being deflected from it. However, because the body of the adult is relatively heavy and the drag forces are relatively small, the body attitude appears to remain more or less constant throughout a long jump rather than following the curve of the trajectory (H. C. Bennet-Clark, unpublished) and the same is true when the body is either projected or the insect jumps vertically. Thus, the potential for reducing drag may exist but may not be fully realised.

On some of the smaller insects, such as fleas, the hind legs are left extended during the early part of the jump so that, although the body may rotate along any of the three major axes, pitch, yaw and roll (Bennet-Clark & Lucey, 1967), it may be stabilised later by the drag of the extended legs so that the attitude through most of the trajectory is with the horizontal longitudinal plane of the body normal to the direction of travel. At the low Reynolds numbers at which fleas move, such mechanisms may be disadvantageous for various reasons. Firstly, much of the drag will be viscous drag due to the limbs, so it may be advantageous to keep some of these folded – fleas fold their pro- and mesothoracic legs before jumping and *Drosophila* fold all their legs during flight. Secondly, it appears that the advantages that might be gained in a small animal by controlling the attitude of the body in flight may be more than offset by the increased drag of legs when they protrude from the body. Fleas, however, jump by extension of their metathoracic legs and the inevitable drag that this mechanism produces can be reduced only by reduction of the length of the legs, which poses other problems. The optimum design is to make the body spherical, as this minimises viscous drag at low Reynolds numbers, and not to have any legs! In the case of fleas, this design requirement conflicts with that for a body capable of pushing through the hairs of the host and of limbs capable of grasping but, even so, fleas tend to be short bodied and, in some species, almost circular in cross-section.

Flea beetles of the subfamily Halticinae of the Chrysomelidae show an approach towards optimising their body design; *Phyllotreta vittula* weighs about 1 mg, is about 2.3 mm long and has a maximum breadth of 1.1 mm. The body density is about 0.75 mg mm^{-3}, which is dense for a flying insect, and from experiments where the insects were fired from a spring gun, the drag coefficient does not exceed 0.7 which is similar to that found for small wooden models and appreciably less than that found for fleas or locusts. As the insects achieve a long jump of 0.4 m (R. F. Ker unpublished), they achieve both longer ranges and higher efficiencies than fleas. Many of the smaller beetles have nearly spherical or hemispherical bodies – this would be a good design for a leaf-dwelling insect, such as a Coccinellid or Chrysomelid operating at small values of *Re* and exposed to winds from all directions: when exposed to winds, ladybirds (*Coccinella* species) clamp their hemispherical bodies down on the substratum using their legs and are very difficult to dislodge with a puff of air.

Conclusions

Aerodynamic drag is a factor which becomes increasingly important with decreasing size, as the frontal area to mass ratio of an animal increases. Various aspects of drag have been considered elsewhere; Alexander (1971) considers the problem of drag on the body of a small animal that might be blown over by wind and shows how this can account for the sprawling gait of insects and he also discusses some of the factors which determine the shape of bodies at different values of *Re*.

Unfortunately, the optimum design for jumping efficiency may be unsatisfactory in other ways; in the very smallest jumping animals, the spherical shape that gives the least drag may be disadvantageous in other contexts. In larger animals, a design for minimum drag may lead to a terminal velocity, when falling, that is wholly unacceptable (Table 7), and so such animals will tend to jump over flat surfaces – as do grasshoppers and frogs – or, if jumping from a height, to extend their legs during the descent, as do many squirrels and hylid frogs. This, too, may explain why many insects, from grasshoppers to fleas (Buddenbrock & Friedrich, 1932; Bennet-Clark & Lucey, 1967) extend their legs when falling; the amputation of legs leads to a reduction of drag and so their extension should not only lead to descent in a stable attitude but also at a lower speed.

Though the present study highlights the importance of aerodynamic drag in one aspect of insect biology, it may be extended to show that with the smaller insects, drag forces from wind may exceed the weight of the animal and so may affect the distribution and choice of habitats of insects of different shapes and sizes.

This review takes its origin from experimental work undertaken with Dr G. M. Alder of the University of Edinburgh and which is published elsewhere (Bennet-Clark & Alder, 1979). The original catalyst to the approach adopted here was a series of calculations by Professor Alexander (1971). I have since benefited from discussion of the aerodynamic problems with Dr L. Leyton and have made considerable use of information which he has drawn together (Leyton, 1975).

Table 7. *Terminal velocity of various animals falling in air (equation 7); it is assumed that C_D is 1*

Animal	Frontal area per unit mass ($mm^2\ mg^{-1}$)	Terminal velocity ($m\ s^{-1}$)	Terminal velocity ($km\ h^{-1}$)
Flea	3.5	2.2	7.8
Locust 1st instar	1.0	4.0	14.6
Locust 4th instar	0.32	7.1	26
Lesser galago	0.025 (est.)	26	92
Man	0.01 (approx.)	40	146

References

Alexander, R. McN. (1971). *Size and Shape*. London: Edward Arnold.

Becker, H. A. (1959). The effect of shape and Reynolds' Number on drag in the motion of a freely oriented body in an infinite fluid. *Can. J. chem. Eng.* **37**, 85–91.

Bennet-Clark, H. C. (1977). Scale effects in jumping insects. In *Scale Effects in Animal Locomotion*, ed. T. J. Pedley, pp. 185–201. London: Academic Press.

Bennet-Clark, H. C. & Alder, G. M. (1979). The effect of air resistance on the jumping performance of insects. *J. exp. Biol.* (In press.)

Bennet-Clark, H. C. & Lucey, E. C. A. (1967). The jump of the flea: a study of the energetics and a model of the mechanism. *J. exp. Biol.* **47**, 59–76.

Buddenbrock, W. von & Friedrich, H. (1932). Über Fallreflexe von Arthropoden. *Zool. Jb., Allg. Zool. u. Physiol.* **51**, 131–48.

Clarke, K. U. (1957). On the role of the tracheal system in the post-embryonic growth of *Locusta migratoria* L. *Proc. R. entomol. Soc. Lond. A*, **32**, 67–79.

Davies, W. M. (1927). On the tracheal system of Collembola, with special reference to that of *Sminthurus viridis*. *Q. J. microsc. Sci.* **71**, 15–30.

Leyton, L. (1975). *Fluid Behaviour in Biological Systems*. Oxford University Press.

Miller, A. (1950). The internal anatomy and histology of the imago of *Drosophila melanogaster*. In *Biology of Drosophila*, ed. M. Demerec, pp. 420–534. London & New York: Wiley.

Nachtigall, W. (1977). Swimming mechanics and energetics of locomotion in variously sized water beetles – Dytiscidae, body length 2–35 mm. In *Scale Effects in Animal Locomotion*, ed. T. J. Pedley, pp. 269–83, London: Academic Press.

Tucker, V. A. (1977). Scaling and avian flight. In *Scale Effects in Animal Locomotion*, ed. T. J. Pedley, pp. 497–509. London: Academic Press.

Vogel, S. (1965). Studies on the flight performance and aerodynamics of *Drosophila*. Ph.D. Thesis, Harvard University, Cambridge, Mass.

Wigglesworth, V. B. (1935). The regulation of respiration in the flea, *Xenopsylla cheopsis*, Roths. (Pulicidae). *Proc. R. Soc., Ser. B*, **118**, 397–419.

W. NACHTIGALL

Some aspects of the kinematics of wing beat movements in insects

Introduction

To understand the kinematics of wing beat movements of a flying insect one has to know the position of a series of points of the wing surface in space during a whole stroke cycle. There are two reasons why physiologists and aerodynamicists are interested in knowing such data. Firstly, if the wing movements are known, it may be possible, under certain circumstances, to calculate aerodynamic forces or to check aerodynamic theories of small oscillating wings. Secondly wing kinematics are an outstanding, periodically repeated indicator of what is happening inside the oscillating thorax box.

Some typical results

Flight of the blow fly, Calliphora erythrocephala

A blow fly, *Calliphora erythrocephala*, fixed to an aerodynamic balance by the tip of its abdomen and flying in front of a wind tunnel, was investigated by Nachtigall (1966), using a three-table projection technique. Pictures were first redrawn by means of a movement analysis projector. Wing beating and wing rotation components could be demonstrated, the latter especially at the upper and lower reversal phases. By adding the vector of translation and developing the projection in a plane it was shown that there is a backward-directed loop in the upstroke (Fig. 1a). This is due to the fact that the backward movement of the wing during the upstroke is faster than the forward movement of the body. The wing is attacked by the air flow at the lower surface during the downstroke and at the upper surface at the beginning of the upstroke. Half way through down- and upstroke, there are small changes of aerodynamic angles of attack. Assuming a 3:1 ratio of lift and drag which seems to be appropriate at these Reynolds numbers, and assuming steady-state flow conditions (which is certainly only relevant for these small, fast-moving wings during stroke periods where angles of attack

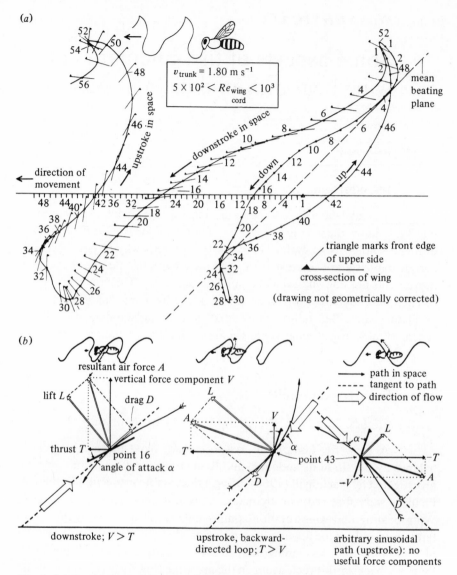

Fig. 1. (*a*) Beating and rotating movements of a *Phormia* wing during
down- and upstroke (not geometrically corrected). (*b*) Sketches of
steady-state aerodynamic forces in the middle of the down- and
upstroke (effects due to non-stationary flow and to wing rotation not
taken into account). After Nachtigall (1966.)

do not change much, i.e. in the middle of the downstroke and first part of the upstroke) one can try some simple aerodynamic constructions (Fig. 1*b*). From these it seems reasonable to assume that there is lift and thrust during both halves of the wing stroke cycle. During downstroke, lift is greater than thrust (Fig 1*b*, left), during upstroke (first part) thrust is greater than lift (Fig. 1*b*, middle). If there was no backward-directed loop during the upstroke, useful aerodynamic force components could hardly be generated (Fig. 1*b*, right). For non-stationary effects see Nachtigall (1979).

Ascending flight of Tipula sp.

Using conventional two-dimensional high-speed film and a method of measuring the changing outlines of wing projections in a horizontal plane Nachtigall analysed the ascending flight of a large dipteran, *Tipula* sp. (Fig. 2*a,b*). It could be shown that the typical coordination of beating and rotation movements of the wings exists in much the same way during ascending flight, in a large, slowly beating (*Tipula*, 50 Hz) dipteran as in a small dipteran with high wing beat frequency (*Calliphora*, 200 Hz). After a few stroke cycles the wings start to beat in an approximately horizontal plane. Wing rotation is maximal at the upper and lower reversal points. The wing is attacked by the air flow at the lower surface during forward movement and at the upper surface during backward movement. Aerodynamic angles of attack are astonishingly high, up to 45° (Fig. 2*c*). Steady-state polars of uniformly attacked *Tipula* wings show that lift is maximal at those high angles of attack; the polars show a broad maximum around 40° (Fig. 2*d*) (Nachtigall, 1977). One must be careful in drawing conclusions, because non-stationary aerodynamic effects may also be important in those larger, slowly beating flies. Furthermore, steady-state polars under uniform flow conditions do not account for rotation movements. Therefore it is also necessary to analyse original *Tipula* wings by moving them like a propeller. Somewhere between these two experimental conditions lies the situation of the beating movement in a living insect, but this situation cannot be analysed directly by experimental methods.

Correlation of kinematics and neuromuscular output in flying locusts

Using the moving-coil method developed by Koch (1977), Zarnack (1978a, b) and Zarnack & Möhl (1977) correlated wing position to neuromuscular events. The locust flew in front of a laminar wind tunnel. I would like to consider only one of a series of detectable effects. The locust was turned slowly to and fro around the dorso-ventral axis (Fig. 3*a*) with an

Fig. 2. Ascending flight of *Tipula*. (*a*) Method of projection. (*b*) Force components at a certain stroke point. (*c*) Beating and rotation movements of a *Tipula* wing immediately after start. (*d*) Steady-state polar of a *Tipula* wing and a flat, thin-wing model in parallel flow. As the polars nearly coincide, the aerodynamical characteristics of the wing are mainly due to its geometrical shape and the Reynolds number. After Nachtigall (1977 and 1979).

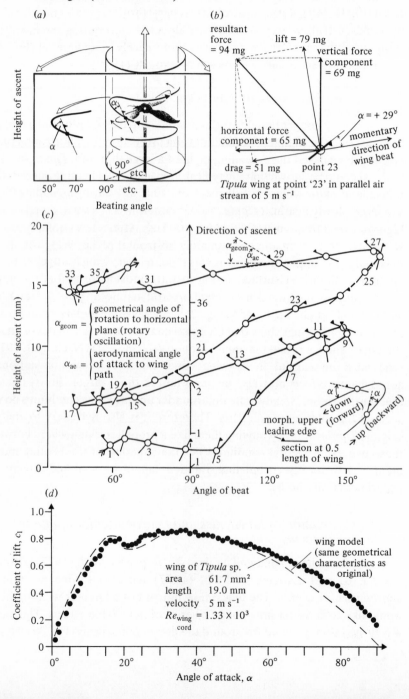

(*a*)

(*b*)

resultant force = 94 mg

lift = 79 mg

vertical force component = 69 mg

$\alpha = +29°$

momentary direction of wing beat

horizontal force component = 65 mg

drag = 51 mg   point 23

Tipula wing at point '23' in parallel air stream of 5 m s⁻¹

Height of ascent

50° 70° 90° etc.

90° etc.

Beating angle

(*c*)

Direction of ascent

α_{geom} α_{ae}

$\alpha_{geom} = \begin{cases} \text{geometrical angle of} \\ \text{rotation to horizontal} \\ \text{plane (rotary} \\ \text{oscillation)} \end{cases}$

$\alpha_{ae} = \begin{cases} \text{aerodynamical angle} \\ \text{of attack to wing} \\ \text{path} \end{cases}$

Height of ascent (mm)

morph. upper leading edge

down (forward) up (backward)

section at 0.5 length of wing

Angle of beat

(*d*)

wing model (same geometrical characteristics as original)

wing of *Tipula* sp.
area   61.7 mm²
length   19.0 mm
velocity   5 m s⁻¹
$Re_{wing} = 1.33 \times 10^3$
cord

Coefficient of lift, c_l

Angle of attack, α

amplitude of approximately 10°. There were detectable changes in the kinematics of the left and right wing which were greater if the angle between the animal's longitudinal axis and the direction of the wind was greater. Fig. 3b shows a typical result. On the upper traces the x, y and z components of the four recording coils are shown as a function of time during the turning movement of the insect around its dorso-ventral axis, one second to the right, one second to the left. Time functions $\triangle t_{x, y, z}$ (explained in Fig. 3c for $\triangle t_x$) are plotted, too. Higher values of these functions mean that the right forewing comes earlier than the left one. It is shown that this is the case during a forced turning of the animal to the left. Apparently the animal tries to orient itself parallel to wind direction; probably the change in kinematics means that the right wing is generating smaller thrust than the left wing, so

Fig. 3. (a) Experimental set-up to rotate a locust around its dorso-ventral axis. (b) Records of the positions of the two forewings of a locust during forced rotation movements and right–left differences in momentary wing positions ($\Delta t_{x, y, z}$, compare the text). (c) Definitions of Δt. After Zarnack (1978a).

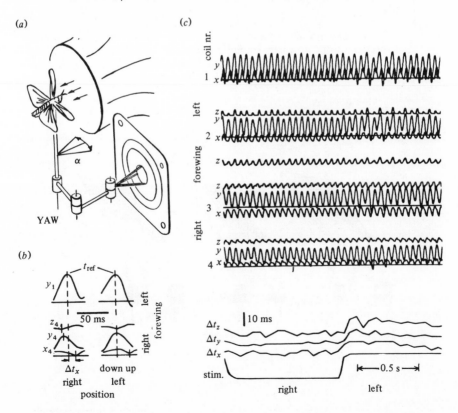

that the animal would turn back parallel to the wind if it was free. According to Pfau's (1977) analysis of functional muscle morphology (Fig. 4) one can assume that, if the right forewing comes later, the right basalar muscles fire later and/or are less active (compare Fig. 4*b* and 4*a*) as soon as the animal is turned to the left. So in the middle of the stroke the aerodynamic angle of attack – i.e. thrust – of the left wing may be relatively higher than in the right wing (Fig. 4) thus providing a torque to restore the animal's original position. However, there is more than just one possibility to change angles of attack. Both basalar and, contrary to earlier findings, subalar muscles provide a downstroke force. Angle of attack depends largely on the relative activity of these muscles. On the other side contractions of a small auxiliary muscle (Snodgrass's no. 85) change the Z-profile of a locust's forewing and its aerodynamic angle of attack, too (Jensen, 1956). So it is not possible to say what really happens inside the thoracic box when stroke differences occur. But synchronous measurements of wing kinematics by the coil method and electrophysiological records of wing muscles, which allow on-line computer correlations, will solve these problems as soon as functional morphology is known in detail.

Fig. 4. Some muscles of a locust's forewing inducing wing beat and wing rotation during downstroke. Different relative activity of these muscles could be responsible for right–left differences in the forewings of obliquely flying locusts; compare Fig. 3*b*. After Pfau (1977).

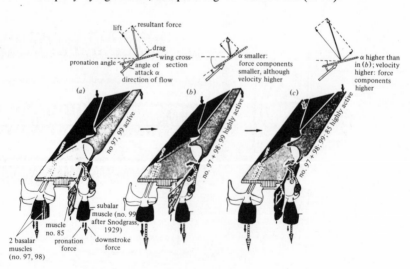

References

Bilo, D. (1971). Flugbiophysik von Kleinvögeln. I. Kinematik und Aerodynamik des Flügelabschlages beim Haussperling (Passer domesticus L.). *Z. vgl. Physiol.* **71**, 382–454.

Jensen, M. (1956). Biology and physics of locust flight. III. The aerodynamics of locust flight. *Phil. Trans. R. Soc. Ser. B*, **239**, 511–52.

Koch, U. T. (1977). A miniature movement detector applied to recording of wingbeat in *Locusta*. Physiology of movement–Biomechanics. *Fortschr. Zool.* **24** (Heft 2/3), 327–32.

Magnan, A. (1934). *Le Vol des Insectes*. Paris: Herman & Cie.

Nachtigall, W. (1966). Die Kinematik der Schlagflügelbewegungen von Dipteren. Methodische und analytische Grundlagen zur Biophysik des Insektenflugs. *Z. vgl. Physiol.* **52**, 155–211.

Nachtigall, W. (1977). Die aerodynamische Polare des Tipula–Flügels und eine Einrichtung zur halbautomatischen Polarenaufnahme. Physiology of movement–Biomechanics. *Fortschr. Zool.* **24** (Heft 2/3), 347–52.

Nachtigall, W. (1979). Insektenflug. Lehrbuchkapital in *Lehrbuch der Zoologie*, begründet von H. Wurmbach, herausgegeban von R. Siewing. (In press.)

Nachtigall, W. (1979). Rasche Richtungsänderungen und Torsionen schwingender Fliegenflügel und Hypothesen über zugeordnete instationäre Strömungseffekte. *J. comp. Physiol.* **133**, 351–5.

Pfau, H. K. (1977). Zur Morphologie und Funktion des Vorder-flügels und Vorderflügelgelenks von Locusta migratoria L. Physiology of movement–Biomechanics. *Fortschr. Zool.* **24** (Heft 2/3), 341–5.

Snodgrass, R. E. (1929). The thoracic mechanism of a grasshopper and its antecedents. *Smithson misc. collect.* **82** (2), 1–111.

Zarnack, W. (1972). Flugbiophysik der Wanderheuschrecke (Locusta migratoria L.) I. Die Bewegung der Vorderflügel. *J. comp. Physiol.* **78**, 356–95.

Zarnack, W. (1978a). Locust flight control. On-line measurements of phase shifting in fore-wing movements. *Naturwissenschaften*, **65**, 64.

Zarnack, W. (1978b). A transducer recording continuously 3-dimensional rotations of biological objects. *J. comp. Physiol.* **126**, 161–8.

Zarnack, W. & Möhl, B. (1977). Changes in the activity of the direct downstroke flight muscles of *Locusta migratoria* L. during steering behaviour in flight. I. Pattern of time shift. *J. comp. Physiol.* **118**, 215–33.

J. M. V. RAYNER

Vorticity and animal flight

Introduction

Natural flight is a subject of great interest to scientists working in many fields. It involves many challenging problems of great complexity, not least because of the many subtle movements made by birds and insects and the great difficulties inherent in observing free-flying animals. The subject may be approached from a number of directions: anatomy, physiology, morphology, energetics, aerodynamics, ecology can all be involved. It is tempting to try to study any one of these aspects of flight in isolation, but this should not be encouraged, for they are all interlinked and we can learn a great deal from studying different aspects together. This paper is concerned largely with the aerodynamics of flight, but it also shows how the aerodynamic model we derive can be a powerful tool in studies of, for instance, avian morphology and flight energetics.

In the discussion that follows it will be assumed that the reader is familiar with the 'biology' of natural flight. There are many publications on this topic to which reference might be made: Pennycuick (1972) is an elementary introduction to many aspects of the subject. A valuable description of the different types of flight and of wing kinematics has been given by Brown (1963); further material of this kind together with fine photographs is in Rüppell (1977). Insect flight is covered by Pringle (1957) and Nachtigall (1974b). Some of these references also discuss flight aerodynamics, which is the subject of valuable reviews by Lighthill (1974, 1977). The aerodynamic properties of wings have been described by Nachtigall (1974a, 1975).

The mechanics of natural flight was one of the many topics to which Leonardo da Vinci turned his attention as part of his research into aeronautics. Since then many distinguished scientists have contributed: among those whose work has been most influential and important are Giovanni Borelli, J. Bell Pettigrew, Etienne Marey and Antoine Magnan. The modern science of aeronautics owes much to the work on bird flight by Sir George

Cayley (*ca* 1805) and to the gliders of Otto Lilienthal based on bird observations (Lilienthal, 1889). The history of the study of natural flight has been discussed in Nachtigall (1973), which contains an extensive bibliography.

Following the examples of Cayley and Lilienthal, aerodynamicists have often looked to natural flight for their inspiration; for instance the recent work on aircraft wing tip sails relates to the separated primary feathers on many bird wings. Now, the study of bird and insect flight calls for many of the theoretical resources of modern aerodynamics before its intricacies can be fully understood. The first author to apply modern wing theory to bird flight seems to have been Walker (1925, 1927); the ideas were taken further by Boel (1929), Stolpe & Zimmer (1937) and von Holst & Küchemann (1941). These last applied a simple two-dimensional *lifting-line* theory to an oscillating rigid wing, and were the first to use Prandtl's and Lanchester's ideas of wake vorticity in the context of natural flight. As we shall see below, no flight can take place without *vorticity* being present in the flow field generated by the wings; vorticity is the fluid-dynamic mechanism which makes aerofoil action, and hence lift and thrust generation by the wings, possible. By using a *blade-element* theory for the wings, Walker and many later authors (see below) have (tacitly, in some cases) acknowledged this fact.

The research which has had the greatest influence on the study of natural flight in recent years is the work on locusts of Weis-Fogh & Jensen (1956). Their painstaking experiment and analysis have led to a much deeper understanding of flapping-wing flight, and are an example to all students of the subject. Weis-Fogh's later work on hovering and unsteady lift generation is also very important (Weis-Fogh, 1972, 1973, 1976, 1977). Other valuable observations of flapping flight are the work of Oehme (see Oehme & Kitzler, 1974, 1975a, b and Oehme, Dathe & Kitzler, 1977) and of Norberg (1975, 1976a, b).

Pennycuick (1968, 1969, 1975) has used the blade-element theory to estimate the aerodynamic power consumption of flying birds, but for a number of reasons we cannot expect the aerodynamic assumptions implicit in that theory to be entirely valid in natural flapping-wing flight. The approach described here (see also Rayner, 1979a, b and c) combines a three-dimensional vortex theory with the frame-work of power calculations used by Pennycuick, in order to derive flight power, and to help us to understand some of the criteria determining a bird's flight style or its morphological character. This is believed to be the first use of a realistic three-dimensional model of the vortex wake of a flying animal.

The flight model

The object of the model described in this paper is to explore the aerodynamic mechanism by which an animal is capable of flight. I shall consider how the wing motions can generate the aerodynamic forces which are required for flight. A flying animal must always contrive to support its own weight; in hovering this is the only force required, but in forward flight it must also overcome the drag of its body and wings. In this paper I shall only consider the forward flight of birds, but much of what I shall say is also relevant to hovering flight and to insects. I shall make the restriction that the bird's upstroke is flexed to minimise aerodynamic drag, and produces no useful aerodynamic forces; this assumption is reasonable for the majority of species (see, for example, Brown, 1963; Norberg, 1975), and in practice only those larger birds that have a tendency to glide or soar are omitted.

From the model we shall be able to estimate the aerodynamic power consumption in various flight conditions (which is the main aim of this research, being of great value to physiologists and zoologists studying energy balance in birds), and ultimately we shall be able to consider the relation between body morphology and power consumption. The saving of energy during flight is of primary importance to a bird, especially when, as in migration, it must fly continually for long periods. From theoretical estimates of power consumption we can use the minimum energy requirement to deduce much about the various forms of adaptation to flight, and about the factors which dictate the choice of a particular flight strategy. In the future it is hoped to widen the discussion by the inclusion of such ecological factors as feeding or breeding efficiency.

The description in this paper is intended largely as an outline of the basic aerodynamics required for the study of natural flight, with a summary of a model which I have recently developed and which incorporates simple aerodynamics with a description of the animal's wing beats to permit power calculations in avian forward flight. A detailed description of the mathematics of the model is given elsewhere (Rayner, 1979b, which discusses hovering, and 1979c, which discusses forward avian flight), while the biological interpretation of the model, together with the ensuing power calculations, is given in Rayner (1979a).

Aerodynamics of a wing

First, I must describe the principles under which any wing operates, since it is important that these are well understood. One point needs to be emphasised at the very beginning: this is the fallacy of the 'rowing' action of a bird's wing (and equally of fish fins or the sails of a yacht). We must dispose firmly of the idea that the propulsive forces generated by the wings are simply the reaction of the motion of fluid (air) displaced by the backward motion of the wings. This idea has in the past been common, and is a very tempting description. However, a glance at the motion of the wings will indicate that it cannot be correct, for the wing is rarely tilted far from the direction of its forward motion, and there could never be sufficient force from fluid reaction to sustain the bird, let alone to overcome the drag of the wings.

The action of the wing may be described by the theory of the aerofoil, which is the basis of much of modern aerodynamics. This helps us to visualise what is happening, but for a number of reasons it will not allow any valid calculations of wing action; a different, but related approach will be invoked.

I have stated above how a wing does *not* work, and shall now attempt to explain how it does. The force experienced by the wing as it moves through the air can be resolved into two components (Fig. 1): the *lift* which is perpendicular to the direction of motion, and the *drag* which is parallel to it. The drag is of course a retarding force, and the bird must do work to overcome it. An aircraft overcomes drag either by the thrust of its engines or by loss of height, but the situation is more complicated in natural flight. The most important difference from aircraft flight is that the bird wing must provide both thrust and weight support by the single wing stroke, so that lift and thrust are part of the same process and cannot be separated. The thrust must include a component to overcome the drag of the beating wings, which cannot be avoided if the wing beat is to produce the necessary forces. While

Fig. 1. Schematic view of the cross-section of an aerofoil and its influence on the incident airflow during steady motion, together with lift and drag forces experienced by the aerofoil.

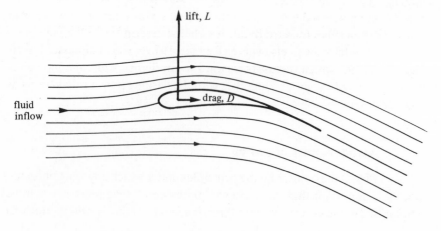

this means that there is some similarity with the action of the rotor blades of a helicopter, the analogy is in general not very helpful. Natural flapping flight is a complex problem which bears little more than superficial resemblance to mechanical flight, for which aerodynamic theory is largely intended.

Wing drag can be resolved into two components, with distinct meanings. *Profile* drag is the result of the form of the aerofoil and of skin friction, and can be expected to be largely independent of the action of the wing. *Induced* drag is the direct detrimental effect of the 'useful' lift generation of the aerofoil, and depends very significantly on the magnitude of the lift. Work must be expended against both of these forces if the bird is to fly; it is possible to think of the work against induced drag as the work required to generate lift. Since the lift is always perpendicular to the aerofoil motion, no work can be done by, or against, it.

Fig. 1 shows how an aerofoil influences the fluid flowing around it, and also the direction of the forces which it experiences. Since the magnitude and direction of the velocity experienced by different portions of a bird wing varies across the span, the magnitude and direction of the local lift and drag forces also varies; for each local cross-section lift and drag will be as shown in Fig. 1, but the total force experienced by the bird will be the vector sum of the forces on the different wing sections. The tapered, cambered cross-section of the aerofoil shown in Fig. 1 is typical of a bird wing, and also of many aircraft. The wing cannot generate any force with a positive component along its local direction of motion. This presents no difficulty to an aircraft, for the engines can produce the thrust required; but, if the oscillating wing of a bird (or insect) is to generate a force in the forward direction which can overcome both wing and body drag then it must beat its wings forwards and downwards in order that the sum of the local lift and drag from the sections of the wing have a substantial forward component (Fig. 2).

Fig. 2. One complete wing beat in forward flight, showing relative lengths of downstroke and upstroke, and *approximate* track of wing tip. Fig. also shows lift and drag experienced *locally* by a wing section during the downstroke.

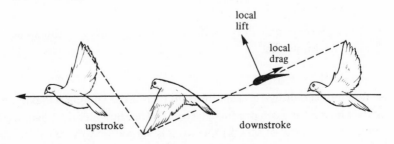

local
lift

local
drag

upstroke

downstroke

During the upstroke the wing is flexed for otherwise both the local lift and drag would have some retarding effect; therefore *all* weight support and thrust is provided by the downstroke. The sum of the local lift *and* drag of the wing sections provides both lift *and* thrust for the entire bird (both wings and body). Note that the drag during the downstroke can have some lifting effect, so that its effect is not entirely detrimental.

Blade-element theory

Blade-element theory, which is the quantification of the aerofoil action illustrated in Fig. 1, has been applied to natural flight in various forms by several authors (e.g. Walker, 1925, 1927; Boel, 1929; Osborne, 1951; Jensen, 1956; Weis-Fogh, 1956; Weis-Fogh & Jensen, 1956; Pennycuick, 1968; Weis-Fogh, 1972, 1973; Lighthill, 1974; Norberg, 1975, 1976a, b). While it can be a useful approach, there are good reasons for supposing that it will not always apply in natural flight, and these are discussed below. In addition, blade-element theory is not entirely appropriate, for it assumes that lift and drag can be separated from each other, and in flapping flight this is not so, for the wing sections experience the two forces together.

For a simple 'classical' wing operating steadily the lift L and drag D are related to the wing velocity u by the familiar equations:

$$L = \tfrac{1}{2} \rho S C_L u^2,$$

and

$$D = \tfrac{1}{2} \rho S C_D u^2,$$

where S is the wing area, ρ is the air density, and C_L and C_D are the coefficients of lift and drag. These coefficients are non-dimensional, and are largely unaffected by change of scale; they are also independent of velocity and are determined mainly by the cross-sectional profile of the aerofoil. The derivation and use of the lift and drag equations are described in any aerodynamics textbook (e.g. Prandtl & Tietjens, 1934; Milne-Thompson, 1958), and I do not need to discuss them in depth. As stated above, the drag coefficient can be divided into two components in the form:

$$C_D = C_{D0} + C_{Di},$$

where C_{D0} is the *profile* drag coefficient and C_{Di} represents the induced drag. Usually we know that

$$C_{Di} \propto C_L^2,$$

so that induced drag rises rapidly as lift increases. There is a maximum value of C_L which a wing can produce; beyond it (or equivalently, beyond a critical value of the angle of attack) the wing is stalled and there can be no lift.

This is the basis of what is known as *blade-element theory*. To apply it to natural flapping flight it seems that we need only introduce an appropriately varying wing velocity and wing area into the above equations, and then use the total work done against drag over a stroke to give the power consumed; this would only require suitable values for C_D and C_L. Unfortunately there are definite suggestions that this procedure is not appropriate: we do not know the values of the coefficients exactly, and there are disturbing indications that the correct values of the coefficients are not consistent with steady-state blade-element theory as it is normally understood. The maximum value of C_L consistent with the theory for a simple wing is about 1.0; in exceptional circumstances it may reach as high as 2.0. Observations of a hovering flycatcher *Ficedula hypoleuca* suggest the value 5.3 (Norberg, 1975); even allowing for experimental and numerical errors this is disturbing. Similar high values were also found for bats (Norberg, 1976a, b). Aerodynamic calculations (Hummel & Möllenstädt, 1977) based on observations of a sparrow also suggest unusually high values in slow flight (*ca* 2.0). While there may well be modifications to a bird's wing which make these high values feasible, we have no knowledge of appropriate values, or of how they may be obtained. It is clearly improper to base estimates of the coefficients on simple steady-state aerofoil theory in our present knowledge: all that we do know is that the bird must have some mechanism permitting far higher values of lift coefficient than can be expected for a simple rigid wing.

A related problem is the effect of unsteady flow around the beating wings. Aerodynamic forces are not generated steadily during flight, and indeed the unsteadiness is an important feature (cf. the Weis-Fogh clap and fling mechanism, relevant to pigeons in slow flight and take off (Lighthill, 1973; Weis-Fogh, 1973, 1976)). Although blade-element theory is designed to describe *steady* motion of the aerofoil, it is also suitable for unsteady motion if the time-scale of the changes is low (quasi-steady). In fast forward avian flight conditions are usually quasi-steady, but in slow flight and hovering they are not (as indicated by the high lift coefficients), and we should not rely on the blade-element theory unless we can be certain of how well it applies. For a further description of this topic, see Rayner (1979c, §5).

The explanation of how high lift coefficients are obtained is undoubtedly related to the unsteadiness of the wing beat, but as yet I can do no more than speculate on how this mechanism might operate. Since the whole topic of the lift coefficient and wing unsteadiness is confused and complicated, it is better to follow the indications that the mathematical formulation of blade-element theory given above is unsatisfactory, and to look elsewhere for a theoretical approach to natural flight aerodynamics.

The main difficulty at this stage is to see how to analyse the wing move-ments and the air motions they create in a way which will permit calculation of the power consumed. I have explained the method in detail elsewhere, and only give the main principles here. The approach is novel (at least in the discussion of flapping flight), and has the great advantage of avoiding many of the features which detract from previous models. The solution is to direct attention away from the wings themselves, and instead to consider the air motions that they generate.

Vorticity

To proceed further we must look more closely at how an aerofoil influences the fluid flowing past it in order to generate lift. We will need to introduce the concept of *vorticity*, which is at the heart of aerodynamics, and without which much animal locomotion in fluids would be impossible; since the main principles are vital to this discussion I shall outline them here.

Vorticity is a measure of the local curvature or rotation of the velocity in a fluid flow. The fluid must be considered as a continuum which conserves volume as it travels (incompressible); this is a reasonable description of most flows in water and air, and therefore the majority of common fluids. At each point in the fluid velocity and vorticity are implicitly related, but it is perfectly possible for a flow to be entirely free of vorticity, or to have vorticity confined to the solid surface which surrounds it. Vorticity does *not* represent the direction of local fluid translation, but rather is related to the change locally in that translation. It is a vector quantity, which is locally perpendicular to the plane containing the greatest fluid rotation (circula-tion). We can think, equivalently, of a vortex line (line defined at every point by the direction of the vorticity vector) as encouraging the fluid to circulate about it. The most important properties of vorticity are that

(i) vortex lines always form closed loops, or end at a surface;
(ii) any distribution of vorticity may be described as an amalgam of vortex loops;
(iii) vorticity is convected with the local fluid velocity;
(iv) vorticity can only be generated at a solid surface bordering or immersed in the fluid.

To get some idea of the meaning of vorticity I mention very briefly some familiar examples. First, the tornado or dust devil (Fig. 3a); this is effectively a thin vortex line stretching from the ground to the clouds, and forcing the air to rotate vigorously around it. This forced circulation is the most important characteristic of a vortex. Similar to the tornado is the bath-tub vortex, or whirlpool (Fig. 3*b*). The fluid circulates, due to Coriolis forces, as it falls into the sink, and a vortex is set up as a result; the dip in the centre is caused by a disparity in pressure between the atmosphere and the rapidly rotating fluid in the core.

These two examples are not at all relevant to natural flight, but they do illustrate a *straight* line vortex in familiar situations. But, not all vortices are straight, and the next example, which is of great importance in animal locomotion, and which will be very familiar, is the *vortex ring* (Fig. 4), commonly known as the smoke ring. Here the vorticity occupies a torus, which is usually circular, and the vortex lines run parallel to the axis of the torus core. This configuration has one important characteristic: once gener- ated it exists and travels apart from any surface; further, if it is circular, it

Fig. 3. Examples of vortex flows. (*a*) Tornado or 'dust devil'; (*b*) bathtub vortex or whirlpool. Double-headed arrows indicate sense of vorticity.

(*a*) (*b*)

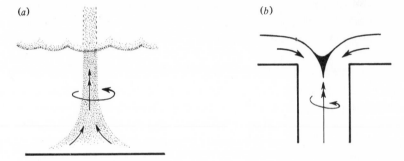

Fig. 4. A typical vortex configuration: the vortex ring. The arrow shows the sense of vorticity in the core of the ring, and the dashed lines show the velocity field induced by the ring in the surrounding fluid.

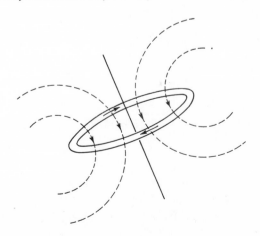

travels for some considerable time without distorting or breaking up. Normally the vortex ring is invisible, but in a smoke ring the core can be seen because it contains tobacco vapour. Vortex rings occur naturally at the top of a rising plume of hot air during a warm day with anticyclonic conditions (a thermal), where they assist many larger birds to gain height by soaring; here again, the vortex is usually invisible.

A simple way to see a vortex ring (or at least a part of one) is to draw the tip of a spoon across the surface of a cup of black coffee. Two small dimples appear just behind the spoon and travel close to it: these are the ends of a semi-circular vortex ring which was generated at the edges of the spoon.

Why does a vortex ring travel, even without any external influence? Fig. 5 shows a cross-section of a circular ring, together with the velocity field that it induces in the surrounding fluid. Evidently the velocity induced on the right-hand side by the left is the same as that induced on the left-hand side by the right: both sides travel downwards, and since the ring is circular the induced velocity is constant around the circumference and the ring travels without growing or distorting. Since the ring is travelling without any external force being applied it must possess momentum as an 'intrinsic' property; further, work must be done to generate the ring, so it will also 'possess' energy. The momentum and energy must be equal to the impulse and energy required to generate the ring from rest in undisturbed fluid, and are equivalently the momentum and energy of the fluid motions induced in the air surrounding the ring.

The sense of the circulation about the ring as drawn in Fig. 5 is such that

Fig. 5. Cross-section of a circular vortex ring, showing the circulation and induced velocity field, and indicating why the ring travels downwards.

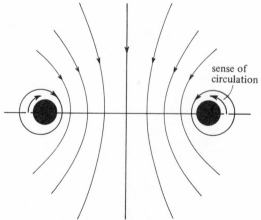

the vorticity vector points out of the page on the right side, and into the page on the left; looking along the axis in the direction of travel of the ring, the vorticity will point clockwise around the ring (see Fig. 4). It is important to recall that since it is a measure of rotation or circulation vorticity possesses a definite parity, which is reflected in the methods by which it is defined and used; the apparent loss of symmetry is no complication! The sense of a vortex ring depends on the method of its generation: it would be possible for the ring shown in Figs. 4 and 5 to have the opposite sense, but it would then have the opposite momentum (although the same energy), and would travel upwards.

Further information about vorticity and its importance in aerodynamics will be found in most fluid-dynamic textbooks (e.g. Batchelor, 1967, where the mathematical formulation is given in detail).

Vorticity and natural flight

Related to the topic of vortex ring momentum is a vital point, of paramount importance in the study of animal locomotion at *high Reynolds number* $(Re \geqslant 10^3)$.

Reynolds number describes the relative importance of viscous (due to viscosity, v) and inertial forces (due to fluid momentum) in a flow. It is defined as $Re = U\,l/v$, where U is a measure of velocity, and l is a typical length scale. The higher the value of Re, the less important the effect of viscosity, and the more the flow is dependent on momentum and vorticity. In aerodynamic situations Re is typically of the order of 10^4 or greater; values between 10^3 and 10^5 are typical of swimming fish or flying birds, and also include flows due to the motion of cetacean mammals, swimming snakes, Cephalopods, and most birds. For the smallest insects and most other invertebrates, and also most animals on a smaller scale, the flow is dominated by viscosity, and vorticity cannot persist; in these low Reynolds number flows motion must be by phenomena such as 'rowing' (e.g. in water-beetles: Nachtigall, this volume, pp. 107–24), or cilia and flagella action.

When the flow is at high Reynolds number viscosity is negligible in the body of the fluid (although it is still important close to surfaces), and motion is dominated by inertia: Newton's third law, that action and reaction are equal and opposite, must always be obeyed. This means that if any animal propels itself through the fluid, be it water or air, it requires momentum as thrust to overcome drag and as lift to overcome body weight, and that therefore it must generate equal and opposite momentum in the fluid flow composing the wake of the animal. *No fluid flow can contain momentum in the absence of an externally imposed force unless there are closed vortices present*; a closed vortex is the only structure with the necessary property of retaining momentum. There are no forces acting on an animal's wake other than the body motions which generate it, and therefore in the wake of every swimming fish, or flying bird, or jetting Cephalopod we shall find a distribution of closed vortices.

Vorticity and wing theory

A wing section (Fig. 1) is cambered in such a way that the fluid flowing past it is deflected downwards: the extra downward momentum imparted to the flow gives the wing lift (and also induced drag) as its reaction. But why is it deflected? Viscosity on the surface of the wing causes vorticity to be generated: this has the effect of accelerating the flow over the upper surface, and decelerating the flow under the lower surface, and the pressure discrepancy between the two surfaces results in lift. The vortex on the wing is similar to, and may be modelled as, a straight line vortex attached to the leading edge of the wing. There cannot be vorticity on the wing without further vorticity being shed into the wake: two trailing straight line vortices are shed from the wing tips, and because vortex lines are closed there must also be a vortex line along the position where the wing started to move (the starting vortex); see Prandtl & Tietjens (1934, pp. 185–206). The vortex line on the wing induces the downward deflexion of the wake.

A simple aerofoil-section wing generates lift by setting up a vortex wake comprising a pair of line vortices. In fact the situation for most wings is slightly more complicated, for lift is not evenly distributed across the wing span; a *vortex sheet* is shed from the trailing edge of the wing (all vorticity is confined to a thin surface). This vortex sheet rolls up rapidly behind the wing into a pair of trailing vortices which are not as widely spaced as the wing span (Milne-Thompson, 1958, pp. 185 and 186). These vortices form the vapour trail visible behind an aircraft; the low fluid pressure in the centre of the vortex core where the rotation is greatest encourages water vapour to condense, and together with the exhaust makes the vortex cores visible.

The vortex theory of animal flight

I have described above the essentials of aerodynamic theory, and shall now discuss how they are relevant to natural flapping flight, and in particular to the forward flight of birds. Theoretically the principles would also apply to ornithopter design, but as an incidental the calculations show clearly that flapping-wing machines of the size necessary to support a man would need far too great a power/weight ratio to be practicable, and that the wings would need to be far too long to remain sufficiently rigid and light if the machine were to take off under its own power.

A bird's wing must leave a vortex sheet as it travels, for it operates exactly like the aerofoil described above. The main distinction is that the flow is unsteady, a direct result of the oscillating wing beat. During the upstroke the wings have no bound vorticity, for they are flexed and generate no aerodynamic forces. At the top of the upstroke the wings are flicked up rapidly to their highest position ready for the start of the downstroke; they then begin to move down rapidly, lift is produced, and a vortex sheet must be formed. The flick generates vorticity of the correct sense on the wings, and therefore reduces the delay while the wing builds up to its maximum lift during the early part of the downstroke. In some species the wings are close together at the top of the downstroke (e.g. pigeons, terns); this proximity will help lift generation by a mechanism related to the Weis-Fogh clap and fling. The wings shed a vortex sheet during the downstroke, and since all of the vorticity is 'shaken off' the wing at the end of the downstroke, the succeeding upstroke can be unloaded and no vorticity is then shed. The elements of vorticity shed by each successive downstroke will be separated

Fig. 6. The vortex distribution generated by a single downstroke. (*a*) The vortex sheet shed during the first half of the downstroke. Single arrows indicate the sense of the vorticity in the sheet; double arrows indicate the bound vorticity on the wings. (*b*) The half vortex ring formed by roll-up of the vortex sheet shown in (*a*). The sense of the vorticity is such that the vortex ring is convected downwards. (*c*) Immediately after the start of the upstroke, showing the complete elliptic vortex ring. (*a*) and (*b*) after Rayner (1979a).

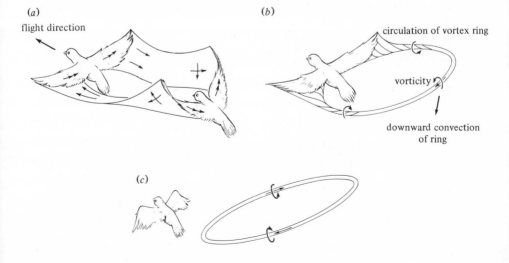

from each other, and are spaced out along the wake owing to the forward motion of the bird and the downward convection of all wake vorticity.

To see how the vortex sheet develops during the stroke we refer to Fig. 6. Fig. 6a shows the vortex sheet, as shed during the first half of the down-stroke; it lies along the trajectory of the wings. The sense of the local vorticity on the sheet and the wings is indicated by arrows. By analogy with the plane vortex sheet behind a rigid wing, this vortex sheet can be expected to roll up rapidly into a line vortex, which must form a portion of a closed loop. Moreover, the twist of the sheet caused by the downward motion of the wings encourages roll-up, and we can expect the vortex configuration at the middle of the downstroke to be the part-formed vortex ring shown in Fig. 6b.

Since the bird is moving forward through the air, the vortex ring will be elongated in the direction of flight; it is actually elliptic rather than circular. The remaining half of the ellipse is formed during the latter half of the downstroke (Fig. 6c). Each downstroke generates a similar vortex ring, and the wake is composed of a chain of rings trailing behind the bird, each inclined to the direction of flight (Fig. 7).

This approach represents a considerable advance over previous arguments, for a number of reasons. Previous models have either been based on the unsatisfactory momentum jet theory, which is really only valid for helicopters, jets or airscrews where wake generation is steady (Pennycuick, 1968; Weis-Fogh, 1972, 1973), or have relied on comparisons with metabolic measurements and unjustified assumptions about muscular effi-

Fig. 7. The chain of vortex rings forming the wake in forward flight. The forces acting on the bird are as follows: $M\mathbf{g}$, weight; $\mathbf{D}_{par}$, parasite drag; $\mathbf{D}_{pro}$, profile drag; $\mathbf{Q}$, reaction of vortex ring momentum. These must be in equilibrium when flight is steady and level.

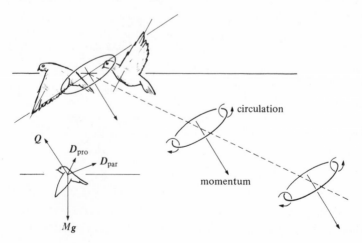

ciency (Tucker, 1973; Greenewalt, 1975). The main advantages of the vortex theory are as follows:

(i) the use of vortex loops is in agreement with how we know a wing operates;

(ii) wake generation is realistically described;

(iii) the unsteadiness of wake generation (caused by the oscillating wing beat and the rapidity of the wing motions) is fully accounted for, and indeed is an integral part of the flight mechanism, for without it closed vortex loops could not be formed;

(iv) lift and induced drag coefficients are not needed in the calculations, and therefore all the uncertainty inherent in their use is avoided;

(v) a wide selection of kinematic and morphologic parameters can be included, so that any animal and its flight style can be fully characterised;

(vi) power may be calculated independently of any assumptions about metabolism, thereby providing independent corroboration for data from metabolic measurements.

The main disadvantage is that most calculations made by using the model require substantial computation, and as yet I have been unable to derive a formula permitting simple use of the theory without a great deal of numerical work.

It should not be thought that these arguments are merely a theoretical indulgence: there are a number of experimental observations which confirm the vortex ring model of the wake. Magnan, Perrilliat-Botonet & Girard (1938) succeeded in photographing 'tourbillons' behind a pigeon in slow flight, using tobacco smoke to make them visible. From Magnan's description and the rather indistinct photographs these seem to be vortex rings. More recently, C. P. Ellington (1978, and private communication) has succeeded in photographing vortex rings and other related structures in the wakes of hovering insects, with talcum powder to visualise the flow field. Fine photographs of vortex rings in the wake of flying finches have been obtained recently by Kokshaysky (1979); these clearly show a chain of vortex elements forming the wake, which are generated in exactly the manner suggested here. To be successful, flow-visualisation experiments on the wake of flying animals require great care and patience, for there are substantial difficulties. It is hoped that future work of this kind will greatly clarify our understanding of wake dynamics.

Power calculations

I must now outline briefly how the model is used to calculate flight power. Each downstroke generates a single elliptic vortex ring, and the wake is composed of a chain of such rings spaced out behind the bird (Fig. 7).

The bird is in *steady*, level forward flight, and the forces acting on it must be in equilibrium: the parasite drag of the body, the profile drag of the wings, and the bird's weight must be balanced by the reaction of vortex ring momentum. From this condition we can calculate the shape and strength of each vortex ring, and therefore the energy required for wake generation. Just as we divide drag into three components, parasite, profile and induced (although in this model induced drag is replaced by part of the reaction of vortex ring momentum), so also may we divide power consumption into the same three components. Parasite power and profile power are simply the rate of working against body and wing drag, and may readily be calculated. Induced power is the rate of working required to generate the vortex wake, and is calculated as the mean rate of increase of wake kinetic energy. A further power component, the inertial power representing the energy lost accelerating the wings, is usually sufficiently small to be neglected in flying birds (Rayner, 1979a, c), but can be important in insects (Weis-Fogh, 1972, 1973); there is no mean 'drag' component due to wing inertia since the downstroke acceleration and deceleration are approximately symmetric.

As forward speed increases, parasite and profile powers both rise, approximately as the cube of flight speed, while induced power falls. For any individual bird we may plot power against velocity as the U-shaped curve shown in Fig. 8. This shape is familiar (it was first found by Pennycuick, 1968, and has been widely used since then), but in general the curve as we calculate it has a rather deeper minimum: Pennycuick interpreted profile power as approximately constant, while in fact it rises considerably (the results are not directly comparable, for Pennycuick used measured, varying, values for the kinematic parameters while we keep them constant (Rayner, 1979a)). Fig. 8 illustrates the $P(V)$ curve for a pigeon *Columba livia* of mass 0.333 kg and wing span 0.632 m, with stroke period 0.15 s. By studying such graphs for different choices of kinematic parameters (e.g. downstroke amplitude, duration, stroke period) we can deduce much about the choice of flight style which the bird should adopt (for details see Rayner, 1979a, §4), and indeed the strategy which we predict as that which minimises power consumption agrees well with observations of actual behaviour. This strategy is to use as fast and as large a downstroke in slow flight as possible, changing to a slow, small downstroke at about the minimum power speed. Some birds can adjust to the limit of effectively no downstroke, which is gliding (or soaring), but many cannot obtain sufficient lift in this way and must retain some active downstroke. (In gliding the energy required for forward motion can only be obtained at the expense of loss of height, so the strategies are not strictly comparable.)

The speed and amplitude of the downstroke must remain within the limits determined by muscle strength, wing inertia, and the need to provide sufficient lift and thrust; moreover, if flight is continuous over a long period, power consumption must lie within the range in which the muscles work aerobically. These, and other similar, constraints limit the bird's choice of flight strategy to a far narrower range than we might theoretically consider to be possible. The quantification of these limits remains a major outstanding problem; with more data this should not present much difficulty, and it will greatly extend the efficacy of the model for discussion of flight style or of the relation between morphological characteristics and flight behaviour.

These calculations give an estimate of the *aerodynamic* power required for flight. The actual cost to the animal (and that it must count in determining flight strategy) is considerably greater, since energy is lost in the chemical reactions in the muscles and in the mechanical linkages of bones, muscles and tendons. Some estimates put the efficiency as low as 20% (Tucker, 1973), so that the rate of depletion of the bird's energy reserves could be up to five times greater than estimated aerodynamic power. Several authors have measured metabolic power in different conditions by a variety of methods (see reviews in Tucker, 1973 and Greenewalt, 1975; also

Fig. 8. U-shaped curve of aerodynamic power consumption plotted against flight velocity for a pigeon *Columba livia* (mass $M = 0.333$ kg). Minimum power speed is about 11.7 m s^{-1}, cost of transport at this speed is about 0.24, but it does not vary significantly over the speed range $10-17 \text{ m s}^{-1}$. After Rayner (1979a).

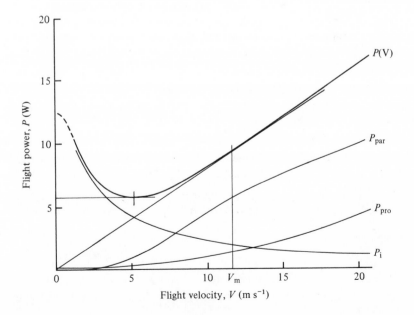

Torre-Bueno & Larochelle, 1978), but until independently determined values for the efficiency are available these metabolic observations cannot be compared with aerodynamic theory. There is reason to believe that efficiency depends on flight velocity and wing beat kinematics, but as yet there is no exact suggestion of how. Those comparisons between theory and experiment which have been made are far from convincing (Tucker, 1973), but the problems involved are considerable, and there are many sources of uncertainty and error. Until a more exact determination of efficiency is available, such comparisons can only be misleading, and cannot be encouraged; however, if efficiency is approximated as lying in the range 25–40%, the order of magnitude of power calculations with this model is confirmed.

The $P(V)$ curve allows us to define a characteristic speed V_m which is of great significance to the animal; this is the speed at which the *cost of transport* is minimum. Cost of transport $C(V)$ is the energy required to carry unit weight of the bird through unit distance (see, for example, Greenewalt, 1975), and is given by the relation

$$C(V) = P(V)/Mg\ V,$$

where M is the animal's total mass and g is the acceleration due to gravity (Mg is the animal's weight). On a long flight where distance rather than duration is important, the bird should seek to minimise $C(V)$, and therefore should fly at about the speed V_m where the tangent from the origin meets the $P(V)$ curve. Cost of transport is perhaps the best indicator of a bird's flight proficiency: higher values mean higher energy consumption, and unless the bird has large flight muscles (and hence more power available) or large fat reserves (and hence more energy) its range might be significantly limited. This is the case for instance in the wren *Troglodytes troglodytes*, which has very small flight muscles (Greenewalt, 1962), and may well only be able to fly with its muscles anaerobic (Rayner, 1979a).

Data for morphologic parameters are available in the literature (for instance, Greenewalt, 1962, quotes wing area, wing span, muscle mass and total mass, among others, from Magnan, 1922), and these can be substituted in the model to calculate flight velocity, power consumption and cost of transport for a wide selection of species. Fig. 9 shows the minimum cost of transport ($C(V_m)$) plotted against body mass (M) on a logarithmic scale for about forty-five species, together with the line of best fit, which corresponds to the allometric relation

$$C(V_m) = 0.212\ (M/\text{kg})^{-0.07}.$$

Cost decreases slightly as mass increases; this is a direct result of the low increase in disc loading (body weight divided by wing span squared) with body mass for otherwise similar species (Greenewalt, 1975; Rayner, 1979a), which is a consequence of the longer wings necessary in larger birds for otherwise take-off would be impossible. Although larger species generally have lower cost owing to their lower induced power, they generally have to pay for this by relatively low flight speeds.

The different symbols in Fig. 9 correspond to the division of species into three groups made by Greenewalt (1975); this division was based more on superficial similarity than on taxonomy. Birds are divided into three groups: *passeriformes* (passerines, petrels, herons, raptors, gulls, terns), *shorebirds* (geese, waders, doves) and *ducks* (ducks, divers, grebes); there is a marked rise in disc loading as we pass through the three groups, and this parameter forms the main defining characteristic. There is a corresponding increase in cost of transport, and in general the passerines are the 'most efficient' and the ducks the 'least efficient' in terms of distance travelled for energy consumed; from other adaptations many members of the 'duck' model have unusually high flight speeds. The main exceptions to this rule are some of the smaller passerine birds, which have disproportionately large wings (tits, finches, flycatchers, wagtails, but not wrens, larks, or swifts and swallows); as will be seen from Fig. 9 these all have large cost of transport (they fall well above the regression line), as a direct result of the high profile drag they would experience in fast forward flight owing to their large wings. These species have evolved an effective answer to the problem: they fold their wings for about half of the time that they are flying, and flap slightly harder for the rest of the time. By this strategy, which is the *bounding* flight characteristic of many small passerines up to and including the green woodpecker, the bird can obtain considerable saving in power consumption, and hence in cost, in fast forward flight, while still having sufficient power in

Fig. 9. Calculated cost of transport at minimum flight speed ($C(V_m)$) plotted against body mass (M) on logarithmic scales. Symbols indicate type of bird, after Greenewalt (1975): x, passeriforms; o, shorebirds; •, ducks. Straight line is the best fit for this data: $C = 0.212 \, (M/\text{kg})^{-0.07}$. See text for description. After Rayner (1979a).

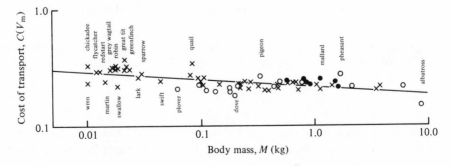

reserve for take off and slow flight when induced power is large. It is interesting that the wren with its very short wings (and small muscles) does not migrate, while the swallows have long thin wings and reduce profile power in that way. All of the passerines that fall above the regression line in Fig. 9 and are of mass less than about 0.25 kg can be observed to use bounding flight for at least some of the time. Bounding flight would be of little value to larger birds, where parasite drag dominates, but some birds (e.g. crows, gulls) can use *undulating* flight, alternating periods of flapping and gliding, to the same advantage.

We may make many deductions of this kind from the model, but at present are strictly limited by the amount of reliable data, both for kinematic and morphologic parameters, which are available. These ideas are expanded elsewhere (Rayner, 1979a), and space does not permit the topic to be discussed in any greater length.

Discussion

The vortex ring model described in this paper seems to be an accurate description of the mechanism by which a bird can fly. There are still a number of points in it that require attention and clarification, but it does allow calculations of power consumption, and helps us to understand exactly why a bird may have evolved a particular shape, or why it chooses a particular style of flight. I have discussed elsewhere the mathematics of the forward flight model (Rayner, 1979c), and considered in detail the conclusions and implications for 'behavioural' studies of flight (Rayner, 1979a, §§ 3–5). The model also describes hovering in both birds and insects; in hovering the wake is axisymmetric, and we evaluate its evolution with time to find the induced power exactly. The calculations are involved, but show clearly that the 'momentum jet' estimate of induced power that is normally used (Pennycuick, 1968; Weis-Fogh, 1972, 1973, for example) is considerably too low for all but the smallest insects, and also that the majority of birds (if not all) do not have sufficient power available for sustained hovering (Rayner, 1979a, §1, 1979b).

As in any model of this kind, this theory relies heavily on observational data for calculations to be made realistically. Although there are many data available in the literature, we cannot rely fully on them, for there are frequent inaccuracies and inconsistencies in the methods of measurement; few taxonomists have been interested in aerodynamics. Future work will include the accumulation of fresh data in a suitable form for aerodynamic studies. It is hoped to investigate the vortex wake experimentally, and in addition there remain a number of problems which demand further attention, including the application of the vortex theory to marine propulsion, of fish, cetacean mammals and cephalopods; as was argued above we must always expect to find closed vortices in the wake of animals moving through fluid at high Reynolds number. Apart from the benefit to physiological and behavioural work, study of the fluid dynamics of locomotion helps us to understand better the evolutionary forces shaping the development of animal species.

References

Batchelor, G. K. (1967). *An Introduction to Fluid Dynamics*. Cambridge University Press.

Boel, M. (1929). Scientific studies of natural flight. *Trans. Am. Soc. mech. Eng.* **51**, 217–42.

Brown, R. H. J. (1963). The flight of birds. *Biol. Rev.* **38**, 460–89.

Ellington, C. P. (1978). The aerodynamics of normal hovering flight. In *Comparative Physiology – Water, Ions and Fluid Mechanics*, ed. K. Schmidt-Nielsen, L. Bolis & S. H. P. Maddrell, pp. 327–45. Cambridge University Press.

Greenewalt, C. H. (1962). Dimensional relationships for flying animals. *Smithson. misc. Collect.* **144**, no. 2.

Greenewalt, C. H. (1975). The flight of birds. *Trans. Am. philos. Soc.* **65**, no. 4.

von Holst, E. & Küchemann, D. (1941). Biologische und aerodynamische Probleme des Tierfluges. *Naturwissenschaften*, **29**, 348–62.

Hummel, D. & Möllenstädt, W. (1977). On the calculation of the aerodynamic forces acting on a house sparrow (*Passer domesticus*) during downstroke by means of aerodynamic theory. *Fortschr. Zool.* **24**, 235–56.

Jensen, M. (1956). Biology and physics of locust flight. III. The aerodynamics of locust flight. *Phil. Trans. R. Soc., Ser. B*, **239**, 511–52.

Kokshaysky, N. V. (1979). Tracing the wake of a flying bird. *Nature, Lond.* **279**, 146–8.

Lighthill, M. J. (1973). On the Weis-Fogh mechanism of lift generation. *J. Fluid Mech.* **60**, 1–17.

Lighthill, M. J. (1974). Aerodynamic aspects of animal flight. *Fifth fluid science lecture, Br. Hydromech. Res. Assn.* Reprinted in *Mathematical biofluiddynamics*. Philadelphia: S.I.A.M. (1975).

Lighthill, M. J. (1977). Introduction to the scaling of aerial locomotion. In *Scale Effects in Animal Locomotion*, ed. T. J. Pedley, pp. 365–404. London: Academic Press.

Lilienthal, O. (1889). *Der Vogelflug als Grundlage der Fliegekunst*. Leipzig: Voigtländer.

Magnan, A. (1922). Les caractéristiques des oiseaux suivant le mode de vol. *Ann. Sci. nat.* (sér. 10) **5**, 125–234.

Magnan, A., Perrilliat-Botonet, C. & Girard, H. (1938). Essais d'enregistrements cinématographiques simultanées dans trois directions perpendiculaires deux à deux de l'écoulement de l'air autour d'un oiseau en vol. *C. r. hebd. Séances Acad. Sci. Paris*, **206**, 462–4.

Milne-Thompson, L. M. (1958). *Theoretical aerodynamics*. New York: Dover.

Nachtigall, W. (1973). Geschichte der Erforschung des Vogelfluges von der Renaissance bis zur Gegenwart. *J. Orn., Lpz.* **114**, 283–304.

Nachtigall W. (1974a). Biophysik des Tierflugs. *Rhein.-Westf. Akad. Wiss. Vorträge*, **236**, 73–152.

Nachtigall, W. (1974b). *Insects in Flight*. London: George Allen & Unwin.

Nachtigall, W. (1975). Vogelflügel und Gleitflug. *J. Orn., Lpz.* **116**, 1–38.

Norberg, U. M. (1975). Hovering flight in the pied flycatcher (*Ficedula hypoleuca*). In *Swimming and Flying in Nature*, ed. T. Y. Wu, C. J. Brokaw & C. Brennen, vol. 2, pp. 869–81. New York: Plenum Press.

Norberg, U. M. (1976a). Aerodynamics, kinematics and energetics of horizontal flight in the long-eared bat *Plecotus auritus*. *J. exp. Biol.* **65**, 179–212.

Norberg, U. M. (1976b). Aerodynamics of hovering flight in the long-eared bat *Plecotus auritus*. *J. exp. Biol.* **65**, 459–70.

Oehme, H., Dathe, H. H. & Kitzler, U. (1977). Research on biophysics and physiology of bird flight. IV. Flight energetics in birds. *Fortschr. Zool.* **24**, 257–73.

Oehme, H. & Kitzler, U. (1974). Untersuchungen zur Flugbiophysik und Flugphysiologie der Vögel. I. Über die Kinematik des Flügelschlages beim unbeschleunigten Horizontalflug. *Zool. Jahrb. Physiol.* **78**, 461–512.

Oehme, H. & Kitzler, U. (1975a). Untersuchungen zur Flugbiophysik und Flugphysiologie der Vögel. II. Zur Geometrie des Vögelflügels. *Zool. Jahrb. Physiol.* **79**, 402–24. Translated as *NASA TT-F*-16901 (1976).

Oehme, H. & Kitzler, U. (1975b). Untersuchungen zur Flugbiophysik und Flugphysiologie der Vögel. III. Die Bestimmung der Muskelleistung beim Kraftflug der Vögel aus kinematischen und morphologischen Daten. *Zool. Jahrb. Physiol.* **79**, 425–58. Translated as *NASA TT-F*-16902 (1976).

Osborne, M. F. M. (1951). Aerodynamics of flapping flight with applications to insects. *J. exp. Biol.* **28**, 221–45.

Pedley, T. J. (ed.) (1977). *Scale Effects in Animal Locomotion*. London: Academic Press.

Pennycuick, C. J. (1968). Power requirements for horizontal flight in the pigeon *Columba livia*. *J. exp. Biol.* **49**, 527–55.

Pennycuick, C. J. (1969). The mechanics of bird migration. *Ibis*, **111**, 525–56.

Pennycuick, C. J. (1972). *Animal Flight*. London: Edward Arnold.

Pennycuick, C. J. (1975). Mechanics of flight. In *Avian Biology*, ed. D. S. Farnell, J. R. King & K. C. Parkes, vol. 5, pp. 1–75. London: Academic Press.

Prandtl, L. & Tietjens, O. G. (1934). *Applied Hydro- and Aeromechanics*. New York: Dover.

Pringle, J. W. S. (1957). *Insect Flight*. Cambridge University Press.

Rayner, J. M. V. (1979a). A new approach to animal flight mechanics. *J. exp. Biol.* **80**, 17–54.

Rayner, J. M. V. (1979b). A vortex theory of animal flight. Part 1. The vortex wake of a hovering animal. *J. Fluid Mech.* **91**, 697–730.

Rayner, J. M. V. (1979c). A vortex theory of animal flight. Part 2. The forward flight of birds. *J. Fluid Mech.* **91**, 731–63.

Rüppell, G. (1977). *Bird Flight*. New York: Van Nostrand-Rheinhold.

Stolpe, M. & Zimmer, K. (1937). Physikalische Grundlage des Vögelfluges. *J. Orn., Lpz.* **85**, 147–64.

Torre-Bueno, J. R. & Larochelle, J. (1978). The metabolic cost of flight in unrestrained birds. *J. exp. Biol.* **75**, 223–9.

Tucker, V. A. (1973). Bird metabolism during flight: evaluation of a theory. *J. exp. Biol.* **58**, 689–709.

Walker, G. T. (1925). The flapping flight of birds. *J. R. aero. Soc.* **29**, 590–4.

Walker, G. T. (1927). The flapping flight of birds. II. *J. R. aero. Soc.* **31**, 337–42.

Weis-Fogh, T. (1956). Biology and physics of locust flight. II. Flight performance of the desert locust (*Schistocerca gregaria*). *Phil. Trans. R. Soc., Ser. B*, **239**, 459–510.

Weis-Fogh, T. (1972). Energetics of hovering flight in hummingbirds and *Drosophila*. *J. exp. Biol.* **56**, 79–104.

Weis-Fogh, T. (1973). Quick estimates of flight fitness in hovering animals, including novel mechanisms for lift production. *J. exp. Biol.* **59**, 169–230.

Weis-Fogh, T. (1976). Energetics and aerodynamics of flapping flight: a synthesis. *Symp. R. ent. Soc. Lond.* **7**, 48–72.

Weis-Fogh, T. (1977). Dimensional analysis of hovering flight. In *Scale Effects in Animal Locomotion*, ed. T. J. Pedley, pp. 405–20. London: Academic Press.

Weis-Fogh, T. & Jensen, M. (1956). Biology and physics of locust flight. II. Basic principles in insect flight. A critical review. *Phil. Trans. R. Soc., Ser. B*, **239**, 415–58.

D. A. DORSETT

Principles underlying the coordination of locomotory behaviour in invertebrates

Introduction

Movement is a fundamental property of all animals. Locomotory behaviour results from orderly sequences of contraction of muscles in the body wall (as in the annelids), or of a number of paired appendages borne on the body segments as in the arthropods.

In annelids the circular and longitudinal muscles are used in various combinations during locomotion. Both alternating and simultaneous use of the circular and longitudinal muscles takes place in the differing forms of peristaltic motion (see Elder, this volume). The dorsal and ventral longitudinals may also combine to form dorso-ventral or lateral swimming waves. Some worms may use more than one of these combinations in different behaviour patterns. In the majority of cases we know little or nothing of the neural mechanisms controlling these events.

The limb movements of many arthropods occur in metachronal sequences forming waves that appear to move anteriorly along the body. The larger crustaceans exhibit a variety of stepping gaits (McMillan, 1975), while the limbs of the insects and arachnids are classically described as forming alternating tripod and tetrapod patterns, based on the tendency of limbs on opposite sides of the body to alternate. The gait of these and other forms of arthropod locomotion such as flight and swimming, can also display the properties of metachrony (Wilson, 1966). Locomotory behaviour is usually sustained for varying periods of time, the limbs and muscles showing *cyclic activity* which is repeated at regular intervals in the *locomotory rhythm*. The implication of these requirements for the nervous system are clear. The neurons in each segment must provide an appropriate pattern of impulses to drive the muscles in their proper sequence, while maintaining the correct phase relationships with the segments on either side. This coordination may be achieved in several ways. The body wall and appendages of the arthropods are equipped with an abundance of sense organs which can provide information on the static and dynamic st te of the limbs and their relationship to the body at any one time. Some r ptors generate reflexes, such as *resistance reflexes*, which activate the muscles opposing the imposed movement (Bush, 1962). It was once considered that locomotory movements were controlled in this way, sensory information generated by completion of one phase of the movement being used to initiate the subsequent stage, the whole cycle being dependent on *chain reflexes*.

No examples are known of simple behaviour patterns which are controlled in this way. In fact, the reverse is the case. The swimmeret movements in the crayfish, the flight pattern in the locust, can both be initiated after all afferent inputs have been eliminated (Ikeda & Wiersma, 1964; Wilson, 1961). Neurons in the completely isolated brain of the mollusc *Tritonia* will reproduce some 40 s of integrated neural activity characteristic of the swimming escape behaviour, following a short electrical stimulus to a peripheral nerve trunk (Dorsett, Willows & Hoyle, 1969, 1973). In these organisms the instructions for a pattern of motor impulses generating a properly sequenced series of movements are programmed within the central nervous system in the form of a *motor tape* or *motor score* (Hoyle, 1964). Once the programme is initiated it runs to completion independently of the sensory input it generates. Such behaviour is inherited, resulting from interactions within a set of genetically determined neural connections.

Between the extremes of a *closed loop* chain reflex control system, and the inflexible *open loop* endogenous programmes, there exists a number of intermediate forms of neural organisation. In several instances the centrally generated motor score recorded from de-afferented preparations differs significantly in detail to the motor output from the intact system. For example, the change in impulse frequency observed during the elevator discharge to the leg during the cockroach stepping cycle is reversed by de-afferentation (Pearson & Iles, 1970), and the phase relationship between the dorsal and ventral musculature is fundamentally altered in the swimming leech (Kristan, Stent & Ort, 1974).

The apparently paradoxical situation where resistance reflexes are overridden during centrally commanded motor activity requires some explanation. One suggestion is that the central nervous system contains a *sensory tape* or *template* of the sensory pattern learned from previously successful movements. The incoming activity resulting from a central command is then continuously compared to the expected pattern, and the motor output is adjusted to match or cancel any difference. Difference detection is known to operate during postural control of the abdominal segments of the lobster

(Page & Sokolove, 1972), but how the sensory template is stored in the nervous system is less certain. Support for this type of control mechanism comes from experiments with walking animals. Where the normal sequence is interrupted by some unexpected feature such as a slip, resistance reflexes are then recorded from the muscles to compensate (Barnes, 1977). Many workers have sought to investigate the central mechanisms coordinating limb movements by experiments involving varying degrees of disruption on the peripheral system and studying the consequent effects on the locomotory output. These manipulations have provided much useful information on the ability of the nervous system to compensate for the deficiencies, but reveal no fundamental details of the underlying mechanisms. Such details can be obtained by intracellular recordings made from identified elements in the neural systems generating the behaviour, so that the impact of sensory and central influences can be quantified at all levels.

In the last ten years an increasing proportion of research effort has been directed to this end, but we are still at a stage where attempts to formulate a set of rules or a general synthesis are premature. We have a number of jigsaw puzzles, each with several essential pieces missing, which prevents us from recognising the subject, although the salient features tempt us to guess the identity.

Escape responses – open loop systems

Escape behaviour often falls into a special category with several particular requirements. For a short time, all other activities are subordinated to a single explosive event, or series of events, aimed to remove the animal from a situation which may threaten its survival. It is usual for the behaviour to require a specific sensory modality or input pattern to overcome a relatively high threshold needed for its initiation, and through conducting pathways offering the minimum delay in the effector response.

The tail flip in crayfish

The escape response of crayfish and lobsters provides a good example of the coordination involved in a sudden ballistic movement. The act requires the sudden coordinated contraction of a specialised set of fast flexor muscle fibres in the abdominal segments which propel the animal rapidly backwards and out of possible danger. The muscles are innervated by a special set of fast flexor motoneurons in each abdominal ganglion. One pair of giant motoneurons innervate all the fast flexor muscle fibres in the segment, the remaining eight pairs of non-giant motoneurons innervating smaller but overlapping groups of fibres.

The fast flexor motoneurons are activated by electrotonic synapses formed with the paired median and lateral giant fibres, which are multi-segmental interneurons running the length of the cord (Furshpan & Potter, 1959; Zucker, Kennedy & Selverston, 1971). The two sets of giant fibres mediate quite distinct behavioural responses. The median giants respond to visual and tactile stimuli directed towards the head, and activate the fast flexor motoneurons in all the abdominal segments. This results in complete tail flexion and a horizontally directed backwards movement. The lateral giants, which respond to tactile stimulation on the abdomen, do not synapse with the motor giant neurons in the last two abdominal segments, and by their action on the uropods produce a backwards and upwardly directed movement (Larimer, Eggleston, Masukawa & Kennedy, 1971; Wine & Krasne, 1972).

The tactile sensory input projects centrally on to three sensory interneurons through labile, chemically transmitting synapses, and also directly on to the lateral giant fibres (Fig. 1). The sensory interneurons provide a suitable threshold by intergrating the sensory input over a wide area, and by electrical junctions formed between themselves. The chemically transmitting synapses between the afferents and the interneurons are extremely labile, stimuli occurring as little as once every five minutes leading to a reduction in post-synaptic response lasting several hours. As the hairs are strongly excited during the tail flip itself and during more prolonged swimming sequences, it is obviously necessary to protect the escape reflex from these effects (Schrameck, 1970; Bowerman & Larimer, 1974a, b). During the tail flip the afferent terminals on the sensory interneurons are inhibited presynaptically by a *corollary discharge* from the tail flip motor circuits. This reduces the amount of transmitter released and protects them from the effects of habituation. The corollary discharge also exerts orthodromic inhibitory effects on the primary interneurons and the lateral giant fibres (Roberts, 1968; Krasne & Bryan, 1973), reducing their excitability immediately following the giant impulse.

This example of central inhibition of sensory input generated by a voluntary movement, could illustrate one way in which resistance reflexes are overcome during normal locomotion. The lobster has a reflex which could interfere with the execution of the tail flip. The muscle receptor organs in the abdominal segments operate resistance reflexes which activate the slow extensor musculature (Page & Sokolove, 1972). This could interfere with full abdominal flexion but generally the movement is completed before the sensory information can be utilised by the central nervous system, and central pathways may also be used to inhibit the reflex.

The neural pathways in Fig. 1 show how a simple response involving a four-stage reflex arc and a single effector system is organised. Once the giant fibres are activated the animal is committed to the movement, rapid and effective transmission being ensured by the electrical junctions with the motoneurons. There are no interactions among the motoneurons, each unit being independently recruited by its connections with the giant fibres. This arrangement leaves the motoneurons free to participate independently in behaviour such as swimming, which is controlled by non-giant interneurons.

Fig. 1. Coordinating pathways involved in tail flip circuit of the crayfish. Tactile afferents project to the lateral giant fibre through electrotonic junctions and indirectly by chemically transmitting synapses formed with three sensory interneurons A, B and C. The impulse in the lateral giant activates the giant motoneuron (GMN) and the eight fast flexor motoneurons through electrotonic junctions. A corollary discharge from the motoneuron circuit inhibits the afferent terminals on the sensory interneurons presynaptically (p), protecting them from habituation effects through reflex excitation caused by the movement. From data by Zucker *et al.* (1971); Krasne & Bryan (1973.)

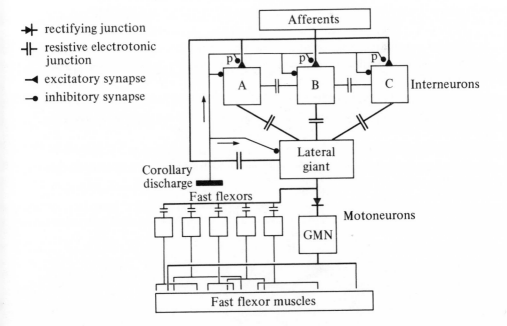

Swimming escape response of Tritonia

Some species of the molluscan genus *Tritonia* undergo a series of stereotyped swimming movements following chemotactile stimulation with tissues from a predatory starfish (Willows & Dorsett, 1975). The behaviour involves a series of alternating and maintained postural changes, involving several distinct muscle systems in different parts of the body. Following the initial reflexive withdrawal from the stimulus the body elongates, forcing blood into the extremities so that the oral veil and tail form flattened, paddle-like structures. The branchial tufts and rhinophores are also withdrawn, helping to streamline the body. These postures are maintained throughout the swim. The swimming movements consist of 5–8 alternating flexions of the dorsal and ventral longitudinal musculature about the mid-point of the body, each lasting some 4–5 s. The behavioural sequence is terminated by a series of incomplete dorsal flexions without an intervening ventral flexion.

Once the behaviour is initiated it runs to completion, after which it can be re-initiated by the appropriate stimulus. In *T. diomedea* the response does not habituate, although there is some reduction in the number of swimming cycles in a long series of trials (Willows & Dorsett, 1975). The neural correlates of part of this behaviour were studied by recording intracellularly from neurons in preparations where the brain had been exposed and stabilised. During swimming, patterned bursting was recorded from pedal neurons which fired in phase with the dorsal flexion (DFNs), the ventral flexion (VFNs), or a few that fired in phase with both flexions (GENs). These neurons were assumed to be motor, or immediate pre-motor units.

Proprioceptive input from the periphery appears to play little part in timing the antiphasic activity of the dorsal and ventral flexion neurons (Fig. 2). Patterned neural activity, identical in all respects to that recorded from the semi-intact preparation, can be observed in the same neurons in a brain

Fig. 2. An ethogram of the swimming escape response of *Tritonia diomedea* recorded from a dorsal (upper trace) and ventral (middle trace) flexion neuron in an isolated brain. The sequence was initiated by a short electrical stimulus to a cerebral nerve trunk. Note the terminal sequence of dorsal flexions without an intervening ventral burst.

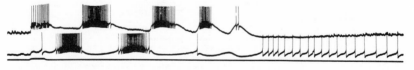

5 s

which has been completely isolated from the animal and stimulated by a short series of pulses to a nerve trunk (Dorsett *et al.*, 1969, 1973). The duration of the response far outlasts the stimulus. The mechanism provides an extreme example of *open loop* control, in which the entire complex sequence is played out by the motoneurons following activation of the *motor tape*.

Details of the neutral interactions which generate the pattern of alternating bursts in the flexion neurons, and time the swimming sequence, have proved surprisingly difficult to obtain. Indirect reciprocal inhibitory pathways can be demonstrated between some pairs of DFNs and VFNs (Fig. 3) and similar excitatory connections were identified between the FNs and the GENs (Willows, Dorsett & Hoyle, 1973; Willows & Dorsett, 1975). These connections provide the basis of an oscillatory network (Csermerly, Harth & Lewis, 1973), although Getting (1977) does not consider feedback from these sources to be a significant part of the pattern-generating mechanism. He has identified a group of primary afferent neurons (the S cells) and several interneurons (the SI group and the C2 cells) which synapse directly with the flexion neurons (Getting, 1976). At the moment it is not possible to

Fig. 3. The current concept of the swimming-escape pattern-generating circuit of *Tritonia*. The alternating flexions are centrally coordinated by the reciprocal inhibitory interactions of the flexion neurons and sustained by the GENs. The flexion neurons receive excitatory and inhibitory inputs from three interneurons one of which (C2) can drive the antiphasic pattern in the motoneurons. The S cells are primary sensory afferents, responding specifically to starfish tissue. Data from Willows *et al.* (1973); Getting (1977).

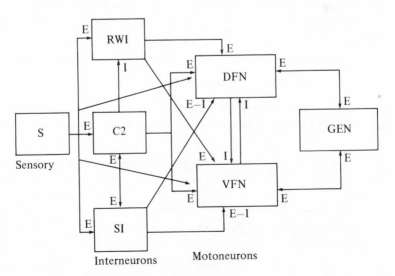

ascribe precise functions to the interneurons, which form both excitatory and inhibitory synapses with the flexion neurons not always correlated with a functional type.

During stimulus-evoked swimming patterns in the isolated brain the interneurons also exhibit patterned bursting, which is not phase-locked to either type of flexion neuron. Despite these inadequacies, tonic activity in the C2 interneuron can result in patterned bursting in the flexion neurons, which may endow it with properties similar to those of command fibres described in the following section. The outstanding feature of the ethogram is the *alternation* of dorsal and ventral motor units. The basis for this reciprocal activity appears to reside in a powerful *inhibitory input* alternately applied to the motoneuron groups, and is a common feature in many locomotory systems.

Sustained locomotory patterns

Most locomotory behaviour involves sustained bouts of activity in the muscular body wall or the limbs, which may last for periods of several minutes. The need for a series of paired appendages to organise into a functional unit calls for the development of neural mechanisms ensuring intra- and intersegmental coordination of the limbs. The most common solution is the *metachronal rhythm* in which the velocity, amplitude and period of the appendages may be continuously adjusted to meet the changing requirements of the organism. How does the nervous system meet these varied demands? Does it need a selection of endogenous programmes to suit each situation, or is a single programme sufficiently flexible to satisfy the requirements?

Swimmeret coordination and command fibres

The swimmerets are paired appendages borne on the ventral side of the abdominal segments in crayfish and lobsters. They beat in a metachronal wave passing anteriorly along the row. The neural mechanisms which underlie swimmeret activity are of particular interest as they are probably applicable to a wide variety of smaller crustaceans which use swimmerets as the principal organs of locomotion. A pair of swimmerets beat in synchrony, the posteriorly directed power stroke alternating with a protraction or recovery stroke. One movement is not a simple reversal of the other, the appendage being folded, twisted and spread by the action of some eleven muscles during the cycle (Davis, 1968). The swimmerets beat as a coordinated series, the power stroke being initiated in the posterior pair and propagating anteriorly with an intersegmental delay of 150–250 ms over the normal range of beat frequencies. Over most of the range the intersegmental phase relationship remains independent of the beat period, with a constant value of around 0.3.

Progressive isolation of the abdominal ganglia from the periphery has little effect on the fundamental motor pattern recorded from the first abdominal nerve roots, which indicates that the burst mechanism is endogenous to the cord (Ikeda & Wiersma, 1964). Cutting the connectives between the last three abdominal ganglia does not impair their ability to generate a patterned output, although the phase relationship between them is lost. Each abdominal ganglion has the capacity to generate a correctly sequenced output of motoneuron activity. When a series of ganglia are connected, the most posterior ganglion dominates, providing the lead for the more anterior ones.

Davis (1969a, b) suggested that the properties of the pattern-generating network in each ganglion (or hemi-ganglion) were best represented by an *oscillator* which provided an alternating drive to the fifty or so motoneurons which operate the swimmeret musculature. The amplitude of the drive is reflected in the intensity of the motoneuron discharge in the abdominal roots, and varies directly with the frequency of the swimmeret beat. As the drive increases, motoneurons are recruited into the burst in order of increasing size, their thresholds conforming to the *size principle* which has been applied to both vertebrate and invertebrate motor systems. The burst duration of individual units is independent of the swimmeret frequency, but the electrical activity recorded in the muscles varies directly with the period. Change in period affects the phase relations between synergistic motoneurons, which overlap as the frequency increases.

Entire patterns of coordinated swimmeret activity can be obtained by stimulating the axons of a number of *command interneurons* isolated from the thoracic regions of the cord (Wiersma & Ikeda, 1964; Davis & Kennedy, 1972). Alterations in the stimulating frequency of individual command units result in relatively minor changes in the period of the swimmeret beat, the impulse frequency of motor units or the number of motoneurons participating in the burst. Some command fibres produce incomplete rhythms, deficient in some aspect of the movement. No single command fibre can reproduce the entire range of periods and beating intensities observed in intact lobsters, and it must be inferred that several command units act in conjunction during the normal beat.

The command units do not carry adequate information for sequencing the swimmeret motoneurons, and as the motoneurons show no evidence of endogenous rhythmic activity, their function must be to excite the pattern-generating networks in the abdominal ganglia (Fig. 4).

Intersegmental coupling is achieved by ascending *coordinating fibres* which carry a *corollary discharge* of the output from the oscillator powerstroke generator, to the oscillator of the next anterior segment. If the signal arrives earlier than expected it advances the phase of the recipient oscillator, if it comes later then the oscillator is delayed (Stein, 1971).

No details have yet emerged of the structure of the oscillator networks, although these are generally assumed to be presynaptic to the motoneurons. It has recently been shown that imposed changes on the activity of some motoneurons can alter the period of the rhythm, which suggests they may form part of, or have access to the oscillator network (Heitler, 1978).

Proprioceptive input from sensory setae and stretch receptors in the coxa provide positive feedback to the powerstroke motoneurons (Davis, 1969b). As the limb velocity declines towards the end of the stroke, the strong inhibitory influence of the setae on the returnstroke motoneurons declines, allowing them to respond to the central drive and the excitatory influence of the coxal proprioceptors. Reciprocal effects are mediated on the inhibitory motoneurons to the two sets of muscles. The proprioceptors may influence

Fig. 4. Coordination of swimmeret movements. Unpatterned discharges from command fibres evoke patterned output from segmental oscillators which drive the swimmeret movements in an alternating fashion. Coordinating interneurons control the metachronal sequence of appendage movements by corollary discharge to the next anterior oscillator. Sensory feedback from the swimmeret modulates the motoneuron output but does not affect the oscillator. After Davis (1969a, b); Davis & Kennedy (1972).

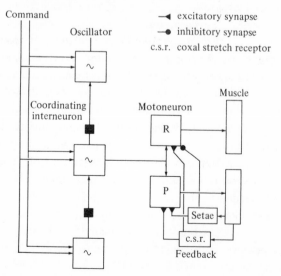

the intensity and period of the motoneuron discharge, but not the phasing or period of the rhythm, indicating their effects are confined to the motoneuron level.

It is now generally accepted that stimulation of single command fibres can evoke behaviour patterns of varying degrees of complexity. Each fibre is unique, although several may contribute to the complete range of a given behaviour pattern. Command units evoking both forwards and backwards walking, swimming and escape behaviour have all been identified from the level of the circum-oesophageal connectives (Bowerman & Larimer, 1974a, b), generally the more complex behaviours being commanded from more rostral regions of the cord. Command neurons do not normally synapse with the motoneurons direct, but act through intermediate networks which fashion the released behaviour.

The swimming leech – a pattern-generating circuit

The identification of the neural elements controlling swimming behaviour in the leech have provided the first detailed account of the mechanism of a locomotory pattern-generating circuit. The properties of the network have many features common to both the swimmeret system (Davis, 1968, 1969a, b) and the pattern generator in the cockroach (Pearson & Iles, 1973). The antiphasic output from two sets of motoneurons provides an alternating drive to reciprocating muscle systems, there is a phase lag between adjacent segments which produces a metachronal wave in the limb or body wall movements, and the ventral nerve cord contains a set of *distributed oscillators* linked by *coordinating fibres* in the connectives (Kristan *et al.*, 1974).

Leeches swim by an undulating wave which passes posteriorly along the body. The period of the wave varies between 400 and 2000 ms, the inter-segmental transmission delay being adjusted so that the body of the leech always occupies one wavelength. This is the optimal shape for hydrodynamic efficiency. The antiphasic contraction of the dorsal and ventral longitudinal muscles is driven by bursts of activity in sets of dorsal and ventral excitatory motoneurons which are symmetrically arranged in each half of the ganglia. Weak electrical junctions link the motoneurons to the synergists of their own side, and to their contralateral homologues. These ensure the motoneurons fire in their correct phase, but avoid jerky movements which might result from impulse synchronisation.

A pair of dorsal and ventral inhibitory motoneurons in each half ganglion synapse with their respective excitatory motoneurons and also inhibit the corresponding longitudinal muscles in the periphery. Reciprocal inhibitory pathways between the inhibitor motoneurons, which form the basis for oscillation in many model networks, were not found in this instance (Ort, Kristan & Stent, 1974). However, stretch-sensitive elements in the body wall generate resistance reflexes which could contribute to the alternation of the dorsal and ventral excitor motoneurons. The output from the stretch receptors disinhibits the excitatory motoneurons to the muscles undergoing stretch, and concurrently excites the inhibitors to the antagonist muscles, whose contraction is causing the stretch (Kristan, 1974). Such interactions could contribute to antiphasic activity of the excitatory groups during swimming, a conclusion which is supported by the finding that following opening the body wall, or deafferentation of the ganglion, the phase-lag between the dorsal and ventral muscles during swimming becomes independent of the period of the wave.

The nerve cord can continue to produce the fundamental rhythm following complete isolation from the periphery (Kristan & Calabrese, 1976), indicating that sensory input is not essential for its generation. In completely isolated cords 8–12 ganglia are necessary to exhibit the swimming rhythm, but if the peripheral connections are left intact, the number of ganglia can be reduced to three. In isolated cords the intersegmental transmission delay is independent of the period of the rhythm, but if a few segments are left intact the stretch reflex can alter the period in more remote segments, indicating

Fig. 5. (a) Intraganglionic connections of the pattern-generating interneurons of the leech. Note the general absence of contralateral interactions, and the inhibitory pathways between the interneurons. (b) The oscillatory loop formed between interneurons 28 and 123 in one half-ganglion and 27, 28 and 123 in the posterior ganglion. From Friesen, Poon & Stent (1978).

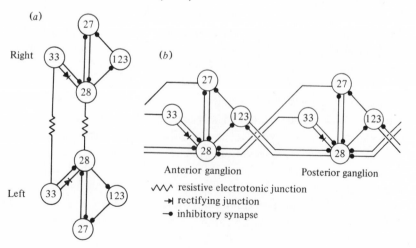

that the sense organs have access to both the motoneurons and the central oscillator, probably through their connections with the dorsal inhibitory neuron.

The basic antiphasic drive to the motoneurons is derived from four pairs of interneurons in each ganglion which establish a network of synaptic connections among themselves and those of adjacent ganglia which is largely ipsilateral and inhibitory. Only two pairs of interneurons in each ganglion make electrical junctions with their contralateral homologues (Fig. 5a). The pattern-generating properties of the network are not immediately obvious from the nature of the established connections, but theoretical considerations and modelling techniques have demonstrated that the network has an inherent rhythmicity based on the principle of *recurrent cyclic*

Fig. 6. Coordination of the motoneuron output by the leech pattern generator. The dorsal (d.e.) and ventral (v.e.) excitor motoneurons and their respective inhibitors (d.i., v.i.) fire in a four-phase cycle. The interneurons 123 and 28 provide the fundamental antiphasic drive to the two excitor groups, 28 supplementing this activity by its opposite action on the two sets of inhibitors. Interneuron 33 excites dorsal excitors in more anterior segments, aided by the action of 27 on the dorsal inhibitors. (Data from Poon *et al.* (1978).)

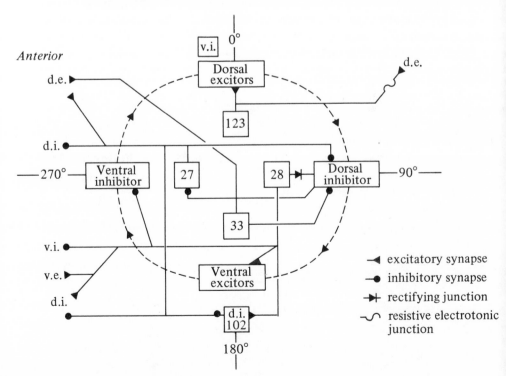

inhibition (Friesen *et al.*, 1978). The fundamental oscillatory loop is formed from an intersegmental chain of two neurons in an anterior ganglion and three in the next posterior ganglion. When provided with constants of the appropriate values, the neurons 28 and 123 alternate in an approximate swimming rhythm (Fig. 5*b*).

Examination of the phase relationships and interactions of the inter-neurons and motoneurons provides an insight into how the mechanism operates (Poon, Friesen & Stent, 1978). The motoneuron activity is based on a four-phase cycle in which the dorsal and ventral excitor motoneurons alternate, and the inhibitors lag their respective excitors by 90° or 180° (Fig. 6). The two interneurons 123 and 28 provide an alternating drive to the dorsal and ventral excitatory motoneurons respectively, and interneuron 28 reinforces the ventral flexion by its connections with the inhibitory motoneurons. The other interneurons in the oscillator loops supplement these activities in more anterior ganglia.

Features of particular interest are the way in which the axons of the interneurons project over several segments, both anteriorly and posteriorly, so that the purely segmental motoneurons receive inputs from more than one oscillator loop. The second feature is the dominant role of inhibitory interactions in determining the output pattern and intersegmental delay. There appears to be no provision for feedback from the excitory motoneurons to the pattern generator, other than by pathways routed through the stretch receptors and the inhibitory motoneurons.

Coordination of flight in the locust

Locusts fly by flapping two pairs of wings borne on the meso- and metathoracic segments. The wings are alternatively elevated and depressed by the coordinated activity of some ten flight muscles, which also control the attitude of the wing during the wingbeat cycle. In many ways the problems of coordination facing the nervous system are similar to those presented by an adjacent pair of swimmerets.

The locust flight system provided one of the earliest examples of a behavioural activity whose motor score was generated endogenously by the nervous system (Wilson, 1961). The flight muscles serving each wing are innervated by some twenty motoneurons whose dendritic and axonal processes are confined to the ipsilateral half of the ganglion (Burrows, 1973b), a whole muscle sometimes being innervated by a single motoneuron. During flight activity an impulse in a wing elevator motoneuron activates a polysynaptic pathway which causes *delayed re-excitation* to itself and a simultaneous depolarisation of its contralateral homologue (Burrows,

1973a). The excitatory post-synaptic potentials (e.p.s.p.'s) causing the excitation are common to both cells, but the amplitude is greater in the contralateral motoneuron and often results in a spike some 40 ms after the original impulse. A similar but separate pathway operates across the ganglion in the reverse direction. A single impulse will occasionally lead to a series of depolarising waves, each delayed by the same interval, suggesting that the pathway is capable of *reverberation*. During flight the two elevator motoneurons of the same segment fire in close synchrony, although they are not electrically coupled, and this leads to summation of the excitatory waves. As the natural flight frequency is around 20 Hz, each elevator motoneuron tends to re-excite its partner one wingbeat cycle later (Fig. 7).

An elevator motoneuron also excites the ipsilateral depressor motoneuron through a pathway which may generate an impulse approximately 25 ms after the elevator spike. This pathway is ipsilateral and one-way, and helps ensure the elevators and depressors of the same side fire in antiphase.

Other interneurons also contribute to the flight pattern generator. Many flight motoneurons undergo regular slow depolarisations in synchrony with the expiratory phase of the ventilatory cycle. Superimposed upon this depolarisation are more rapid oscillations resulting from groups of e.p.s.p.'s

Fig. 7. The flight pattern generator of the locust. The system is unusual in that the elevatoromotoneurons form part of the generator circuit. The delay line and reverberatory circuit are at present unidentified, but probably intraganglionic interneurons. From Burrows (1973a, 1974).

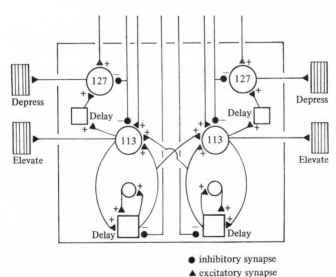

● inhibitory synapse
▲ excitatory synapse

that occur at the periodicity of the wingbeat cycle. The fast and slow rhythms persist in an isolated nervous system and appear to originate in a pair of intersegmental metathoracic interneurons which fire in a burst pattern. The slow rhythm is coded by the burst duration, and the fast flight rhythm by the interval between spikes (Burrows, 1975).

Sense organs on the wing are active during both elevation and depression phases of the cycle (Gettrup, 1963; Wendler, 1974). Raising the wing generates an impulse in the stretch receptor at its base which causes an e.p.s.p. in the depressor motoneurons and an inhibitory post-synaptic potential (i.p.s.p.) in the elevator motoneurons of the same wing. The e.p.s.p. has a short latency and appears to be direct, but the i.p.s.p. is less certain (Burrows, 1974). The forewing stretch receptor also synapses with the ipsilateral depressor motoneurons in the metathoracic ganglion, but does not interact with the elevators, nor the motoneurons of the opposite side. Similarly the hindwing stretch receptors project into the ipsilateral half of the mesothoracic ganglion, where they too only synapse with the depressor motoneurons. Thus raising the wing activates a *negative feedback loop* terminating elevation and helping to initiate depression.

The wing contains a number of sense organs which respond to depression (Burrows, 1976). These elicit responses in the flight motoneurons which are opposite in sign to those generated by the stretch receptor. Usually, the depressor afferents also project to the contralateral motoneurons in the same and adjacent ganglia, and thus have access to the majority of the flight motoneurons. Activation of the hindwing afferents activates the elevator motoneurons of the ipsilateral forewing after a delay of about 10 ms, which represents an appropriate phase shift for normal flight.

At this stage the emerging picture is of a number of systems contributing to the flight pattern generator in the locust. The fundamental oscillatory loop generating the alternating output is unusual in the involvement of the elevator motoneurons as an integral part of the pattern-generating circuit. Other central influences such as command interneurons probably exist to provide excitatory drive to the oscillator, but synaptic input derived from both sensory and interneuron activity has an important influence on the flight 'mood' of the insect by determining the proximity of the motoneuron membrane potential to spike threshold.

An attempt at generalities

In conclusion, I should like to set forth some generalities which may not be universally true, yet seem justified by their occurrence in a number of motor systems.

1. Locomotory sequences are controlled by centrally determined motor programmes.
2. These programmes may be subject to modification:
   (a) by selection of higher-order interneurons, each having a unique set of connections within the pattern generator;
   (b) by the action of sensory feedback on the pattern generator;
   (c) by sensory input modulating motoneuron output.
3. Motoneurons are usually driven by interneurons and not by other motoneurons or by sensory cells. This allows flexibility in motoneuron combinations. Motoneurons of the same class, serving the same muscle or a synergistic muscle may be electrically coupled.
4. Reflexes usually act on motoneurons through interneurons, especially on the contralateral side. Resistance reflexes persist through locomotory sequences but do not control them. Movement detectors are probably less important in this respect than stress or load detectors.
5. Inhibitory effects between interneurons predominate in pattern-generating networks. Impulse activity is not a necessary feature of intraganglionic interactions.

References

Barnes, W. J. (1977). Proprioceptive influences on motor output during walking in crayfish. *J. Physiol., Paris*, **73**, 543–64.

Bowerman, R. F. & Larimer, J. C. (1974a). Command fibres in the circumoesophageal connectives of crayfish. I. Tonic fibres. *J. exp. Biol.* **60**, 95–118.

Bowerman, R. F. & Larimer, J. C. (1974b). Command fibres in the circumoesophageal connectives of crayfish. II. Phasic fibres. *J. exp. Biol.* **60**, 119–34.

Burrows, M. (1973a). The role of delayed excitation in the coordination of some metathoracic flight motoneurons of a locust. *J. comp. Physiol.* **83**, 135–64.

Burrows, M. (1973b). The morphology of an elevator and depressor motoneuron of the hindwing of a locust. *J. comp. Physiol.* **83**, 165–78.

Burrows, M. (1974). Monosynaptic connections between wing stretch receptors and flight motoneurons of the locust. *J. exp. Biol.* **62**, 189–219.

Burrows, M. (1975). Coordinating interneurons of the locust which convey two patterns of motor commands; their connections with the flight motoneurons. *J. exp. Biol.* **63**, 713–33.

Burrows, M. (1976). Neural control of flight in the locust. In *Neural Control of Locomotion*, ed. R. T. Herman, P. Grillner, P. S. Stein & M. Stuart. New York: Plenum Press.

Bush, B. M. H. (1962). Proprioceptive reflexes in the legs of *Carcinus maenas. J. exp. Biol.* **39**, 89–105.

Csermerley, T. J., Harth, E. & Lewis, N.S. (1973). The netlet theory and cooperative phenomena in neural networks. *J. dynam. Syst. Meas. Control*, **95**, 315–20.

Davis, W. J. (1968). The neuromuscular basis of lobster swimmeret beating. *J. exp. Zool.* **168**, 363–78.

Davis, W. J. (1969a). Neural control of swimmeret beating in the lobster. *J. exp. Biol.* **50**, 99–118.

Davis, W. J. (1969b). Reflex organisation in the swimmeret system of the lobster. I. Intra-segmental reflexes. *J. exp. Biol.* **51**, 547–63.

Davis W. J. & Kennedy, D. (1972). Command interneurons controlling swimmeret movements in the lobster. I. Types of effects on motoneurons. *J. Neurophysiol.* **35**, 1–12.

Dorsett, D. A., Willows, A. O. D. & Hoyle, G. (1969). Centrally generated nerve impulse sequences determining swimming behaviour in *Tritonia. Nature, Lond.* **224**, 711–12.

Dorsett, D. A., Willows, A. O. D. & Hoyle, G. (1973). The neural basis of behaviour in *Tritonia*. IV The central origin of a fixed action pattern demonstrated in the isolated brain. *J. Neurobiol.* **3**, 187–300.

Friesen, W. O., Poon, M. & Stent, G. (1978). Neuronal control of swimming in the medicinal leech: IV Identification of a network of oscillatory neurons. *J. exp. Biol.* **75**, 25–44.

Furshpan, E. J. & Potter, D. D. (1959). Transmission at the giant motor synapses of the crayfish. *J. Physiol., Lond.* **145**, 289–325.

Getting, P. (1976). Afferent neurons mediating escape swimming of the marine mollusc. *Tritonia. J. comp. Physiol.* **110**, 271–86.

Getting, P. (1977). Neuronal organization of escape swimming in *Tritonia. J. comp. Physiol.* **121**, 325–42.

Gettrup, E. (1963). Phasic stimulation of a thoracic stretch receptor in locusts. *J. exp. Biol.* **40**, 323–34.

Heitler, W. J. (1978). Coupled neurons are part of the crayfish swimmeret central oscillator. *Nature, Lond.* **275**, 231–3.

Hoyle, G. (1964). Exploration of neuronal mechanisms underlying behavior in insects. In *Neural Theory and Modelling*, ed. R. F. Reiss, pp. 346–76. Stanford, California: Stanford University Press.

Ikeda, K. & Wiersma, C. A. G. (1964). Autogenic rhythmicity in the abdominal ganglia of the crayfish. The control of swimmeret movements. *Comp. Biochem. Physiol.* **12**, 107–15.

Krasne, F. B. & Bryan, J. S. (1973). Habituation: regulation through presynaptic inhibition. *Science, NY*, **182**, 590–2.

Kristan, W. B. (1974). Neuronal control of swimming in the leech. *Amer. Zool.* **14**, 991–1002.

Kristan, W. B. & Calabrese, R. L. (1976). Rhythmic swimming activity in neurons of the isolated nerve cord of the leech. *J. exp. Biol.* **65**, 643–68.

Kristan, W. B., Stent, G. & Ort, C. A. (1974). Neuronal control of swimming in the medicinal leech. 1. Dynamics of the swimming rhythm. *J. comp. Physiol.* **94**, 97–121.

Larimer, J. L., Eggleston, A. C., Masukawa, L. M. & Kennedy, D. (1971). The different connections and motor outputs of lateral and median giant fibres in the crayfish. *J. exp. Biol.* **54**, 391–402.

McMillan, D. L. (1975). A physiological analysis of walking in the American lobster *Homarus americanus*. *Phil. Trans. R. Soc. B*, **270**, 1–59.

Ort, C. A., Kristan, W. B. & Stent, G. (1974). Neuronal control of swimming in the medicinal leech: II Identification and connections of motoneurons. *J. comp. Physiol.* **94**, 121–55.

Page, C. H. & Sokolove, P. G. (1972). Crayfish muscle receptor organ: Role in regulation of postural flexion. *Science, NY,* **175**, 647–50.

Pearson, K. G. & Iles, J. F. (1970). Discharge patterns of coxal levator and depressor motoneurons of the cockroach *Periplaneta*. *J. exp. Biol.* **52**, 139–65.

Pearson, K. G. & Iles, J. F. (1973). Nervous mechanisms underlying intersegmental coordination of leg movements during walking in the cockroach. *J. exp. Biol.* **58**, 725–44.

Poon, M., Friesen, W. O. & Stent, G. (1978). Neuronal control of swimming in the medicinal leech: V. Connections between the oscillatory neurons and the motoneurons. *J. exp. Biol.* **75**, 45–64.

Roberts, A. (1968). Some features of the central coordination of a fast movement in the crayfish. *J. exp. Biol.* **49**, 645–56.

Schrameck, N. E. (1970). Crayfish swimming: Alternating motor output and giant fibre activity. *Science, NY,* **169**, 698–700.

Stein, P. S. G. (1971). Intersegmental coordination of swimmeret motor neuron activity in crayfish. *J. Neurophysiol.* **34**, 310–18.

Wendler, G. (1974). The influence of proprioceptive feedback on locust flight coordination. *J. comp. Physiol.* **88**, 173–200.

Wiersma, C. A. G. & Ikeda, K. (1964). Interneurons commanding swimmeret movements in the crayfish. *Procambarus clarkii. Comp. Biochem. Physiol.* **12**, 509–25.

Willows, A. O. D. & Dorsett, D. A. (1975). Evolution of swimming behaviour in *Tritonia* and its neurophysiological correlates. *J. comp. Physiol.* **100**, 117–34.

Willows, A. O. D., Dorsett, D. A. & Hoyle, G. (1973). The neronal basis of behaviour in *Tritonia*: III. Neuronal mechanism of a fixed action pattern. *J. Neurobiol.* **4**, 255–85.

Wilson, D. M. (1961). The central nervous control of flight in a locust. *J. exp. Biol.* **38**, 417–90.

Wilson, D. M. (1966). Central nervous mechanisms for the generation of rhythmic behaviour in arthropods. In *Nervous and Hormonal Mechanisms of Integration*, ed. G. M. Hughes, *Symposia of the Society for Experimental Biology 20*, pp. 199–228. Cambridge University Press.

Wine, J. J. & Krasne, F. B. (1972). The organization of escape behaviour in the crayfish. *J. exp. Biol.* **56**, 1–18.

Zucker, R. S., Kennedy, D. & Selverston, A. (1971). Neuronal circuit mediating escape responses in crayfish. *Science, NY,* **173**, 645–50.

R. McN. ALEXANDER

The mechanics of walking

Introduction

The slow gaits of mammals and of some birds are generally described as walks. Walking bipeds move their two legs half a cycle out of phase with each other. Quadrupeds move their two forelegs half a cycle out of phase with each other and their two hind legs half a cycle out of phase with each other. Each foot is on the ground for more than half the stride so there are periods when both feet (in bipeds) or both feet of a pair (in quadrupeds) are on the ground.

This paper is about the techniques of walking used by mammals and some birds.

Stiff and compliant walking

Fig. 1 shows four possible techniques of bipedal walking. In the stiff walk (Fig. 1a) each leg is kept straight while its foot is on the ground so that the hip (and the centre of mass of the body) move in a series of arcs of circles. The centre of mass is highest as it passes over the foot. The stiff walk is an

Fig. 1. Diagrams of (a) a stiff walk; (b) another type (i) walk; (c) a level walk and (d) a compliant (type (iv)) walk. Broken lines show the paths of the centre of mass.

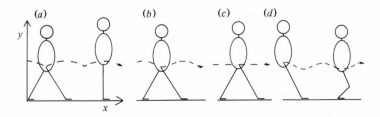

example of the range of walking techniques described by Alexander & Jayes (1978a, b) as type (i) walks. The diagnostic feature of these walks is that the height of the centre of mass has a single maximum in each half stride, occurring as the centre of mass passes over the supporting foot. Another example of a type (i) walk is shown in Fig. 1b. Fig. 1c shows a level walk, in which the legs bend enough to keep the path of the centre of mass level. Fig. 1d shows a compliant or type (iv) walk, in which the height of the centre of mass again has a single maximum in each half stride, but this maximum occurs while both feet are on the ground: there is a minimum while the centre of mass is passing over the foot. Alexander & Jayes (1978a) also defined type (ii) and type (iii) walks, in which the height of the centre of mass has two maxima in each half stride.

The same categories can be used to classify quadrupedal walks if the fore and hind quarters are treated as independent bipeds. Thus the fore legs are performing a type (i) walk if the height of the centre of mass of the fore quarters has a single maximum in each half stride, occurring as it passes over the supporting fore foot.

The normal human walk is a type (i) walk not very different from the stiff walk of Fig. 1a (see Carlsöö, 1972).

There is a maximum speed which cannot be exceeded in the stiff walk (Alexander, 1976). Consider an animal with legs of length h performing a stiff walk with speed u. The centre of mass moves in circular arcs of radius h, implying an acceleration u^2/h towards the centre of the circle (the supporting foot). This acceleration is due to gravity so it cannot exceed the acceleration of free fall, g. Hence

$$u < (gh)^{0.5} \qquad (1)$$

For a typical man walking on earth, $h \simeq 0.9\,m$ and $g = 9.8\,m\,s^{-2}$ so the maximum speed is $3\,m\,s^{-1}$. Men generally break into a run at about this speed (Cavagna, 1969). Athletes walk faster in walking races, at speeds which may exceed $3.7\,m\,s^{-1}$, by a special technique. They keep each leg straight while its foot is on the ground but use a peculiar hip action which flattens the arcs in which the centre of mass moves (Dyson, 1973). This reduces both the amplitude of the vertical movements of the centre of mass and the accelerations required. On the moon $g = 1.6\,m\,s^{-2}$ so the maximum speed for stiff walking is only about $1.2\,m\,s^{-1}$. Astronauts have to use a peculiar loping (slow running) or hopping gait to attain a reasonable speed on the moon (Margaria, 1976).

Equation (1) can be expressed by saying that in stiff walking u^2/gh cannot exceed 1. The parameter u^2/gh is a Froude number, a dimensionless parameter of a type which engineers find useful in discussing motions in which (as in walking) the principal forces are gravitational and inertial. Mammals ranging from small rodents to elephants tend to walk and run in similar fashion at equal Froude numbers: they tend to use the same gait and make stride length the same multiple of leg length (Alexander, Langman & Jayes, 1977; Alexander, 1977b, in which $\hat{u}$ is the square root of a Froude number).

Energy costs: basic principles

The mechanical energy of an animal's body or of any other system is the sum of its kinetic energy, gravitational potential energy and elastic strain energy. These components of mechanical energy are often interconvertible. Energy is converted back and forth between the kinetic and potential forms in a swinging pendulum, between the kinetic and elastic forms in a vibrating spring and between all three forms in a bouncing ball.

The total mechanical energy of the body increases and decreases in the course of a stride. Whenever it increases, the increase must be provided by muscles doing positive work. A decrease may be brought about by air resistance or by frictional or viscous losses in the body. However, the decreases which occur in the course of a walking stride are largely achieved by muscles doing negative work. (Negative work is explained by Goldspink elsewhere in this volume.) Also, muscles may do work against each other: one may do positive work while another does a numerically equal quantity of negative work, so that the total mechanical energy of the body is unchanged. This may occur, for instance, if two feet are on the ground simultaneously, one pushing forward (tending to decelerate the animal) and the other pushing back (tending to accelerate it).

Consider an animal walking at constant speed on level ground. Its stride frequency is f so that a particular stride lasts from time zero to time $1/f$. We will derive an equation giving the metabolic power it needs for walking, making the simplifying assumption that air resistance and other viscous and frictional losses can be ignored. In the short interval from time t to time $(t + \delta t)$, kinetic energy, gravitational potential energy and elastic strain energy increase by δK, δP and δE, respectively. Thus total mechanical energy increases by $(\delta K + \delta P + \delta E)$, which may be either positive or negative, and the muscles must do a corresponding amount of positive or negative work. In the same interval, positive work δW done by some of the muscles is counteracted by negative work $-\delta W$ done by others. The positive work which the muscles must do in each stride is the sum over the time interval 0 to $1/f$, of the work done to increase mechanical energy and the work done against other muscles. It is

$$\sum_{t=0}^{1/f} \{0.5 \mid \delta K + \delta P + \delta E \mid + \delta W\}$$

The vertical lines enclosing the mechanical energy term indicate that this term is to be taken as positive, whether the energy is increasing or decreasing. The factor 0.5 occurs because only half the changes of mechanical energy in a stride are increases: the other half are decreases. The negative work which must be done in each stride has the same numerical value as the positive work. These quantities of positive and negative work must be done f times in unit time. Let η be an efficiency such that performance of one unit of positive work and one unit of negative work requires $1/\eta$ units of metabolic energy. Then the metabolic power required for walking, M, is given by

$$M = (f/\eta) \sum_{t=0}^{1/f} \{0.5 \mid \delta K + \delta P + \delta E \mid + \delta W\} \qquad (2)$$

Let the animal have mass m and let it walk in a system of Cartesian coordinates as indicated in Fig. 1a. At time t its centre of mass has coordinates (x, y). The gravitational potential energy is then mgy. The kinetic energy has several components, of which the largest will generally be the longitudinal external kinetic energy $0.5m(dx/dt)^2$. There is also vertical external kinetic energy $0.5m(dy/dt)^2$. These components of kinetic energy are associated with the movements of the centre of mass but we have to consider in addition the internal kinetic energy due to parts of the body (especially the limbs) moving relative to the centre of mass. Elastic strain energy will occur in any part of the body which is under stress, but especially in the tendons of active muscles (Alexander & Bennet-Clark, 1977; Currey, this volume).

If the forces on the feet were strictly vertical, longitudinal external kinetic energy would remain constant throughout the stride. Force platform records show that the forces slope, keeping more nearly in line with the legs (Fig. 2. The directions of the forces are described more precisely by equation (5).) The force exerted by each foot has a forward component in the first half of the period of contact with the ground, and a backward component in the second half. Consequently the ground exerts a backward (decelerating) force on the body followed by a forward (accelerating) force. This tends to make the longitudinal external kinetic energy fall and then rise again. The statements made in this paragraph apply to all the gaits of all the mammals and birds which have been investigated with force platforms (Alexander, 1977b; Jayes & Alexander, 1978; Cavagna, Heglund & Taylor, 1977).

Fig. 3 shows schematically the fluctuations of mechanical energy which generally occur in type (i) and type (iv) walking. In type (i) walking potential energy fluctuates out of phase with longitudinal external kinetic energy, as for a swinging pendulum. In type (iv) walking elastic strain energy fluctuates out of phase with kinetic plus gravitational potential energy, as for a

Fig. 2. Outlines traced from selected frames of films of a 68-kg man walking and running, and a 36-kg dog trotting, over a force platform. The magnitude and direction of the force recorded at the same instant is shown in each case. (There was only one foot on the force platform in each case.) From Alexander (1977a).

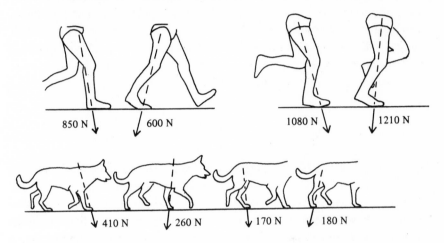

Fig. 3. Schematic graphs of mechanical energy against time for a biped executing (a) a type (i) walk and (b) a type (iv) walk. Total mechanical energy is shown divided into gravitational potential energy (PE) longitudinal and vertical external kinetic energy (KEL and KEV) and (in (b) only) elastic strain energy (EE). Elastic strain energy is ignored in (a) and internal kinetic energy is ignored in both graphs.

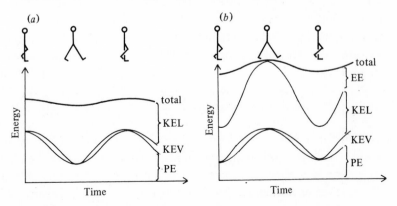

bouncing ball. In both cases the fluctuations of total mechanical energy (and so the metabolic power requirement) may be quite small, but in one case energy is saved by the principle of the pendulum and in the other by that of the bouncing ball (see Cavagna, Heglund & Taylor, 1977).

The pendulum principle makes the type (i) walk economical for bipeds at low speeds (it will be shown that a run is better at high speeds). The same principle could be applied directly to quadrupedal walking if quadrupeds set down their fore and hind feet simultaneously, so that the fore and hind quarters rose and fell simultaneously. The motion of the quadruped would then be just like the motion of two bipeds, walking one behind the other. However, quadrupeds tend to walk with the fore and hind quarters a quarter of a cycle out of phase with each other (McGhee & Frank, 1968; Hildebrand, 1976; Jayes & Alexander, 1978). The type (i) walk is also economical for this gait at low speeds but a different line of argument is needed to demonstrate the point (Alexander, 1980).

Forces on the ground

Fig. 4 shows records of the forces exerted by a man's foot in walking and running. Notice that the records of the vertical component of force differ markedly in shape. This component shows two maxima in walking, and these maxima are separated by a much deeper minimum at the higher speed. There is only one main maximum in running (a small initial maximum due to impact of the foot with the platform is ignored).

Fig. 4. Records of forces exerted on the ground by one foot of a 70-kg man (a) walking at 0.9 m s^{-1}, (b) walking at 1.9 m s^{-1} and (c) running at 3.7 m s^{-1}. F_x is the longitudinal and F_y the vertical component of the force. From Alexander & Jayes (1978a).

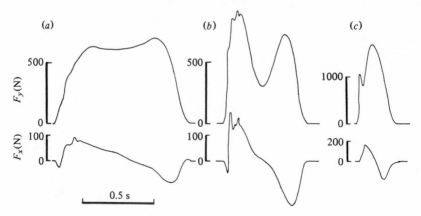

Consider an animal walking with stride frequency f, and duty factor β. This means that each foot is on the ground for a fraction β of the duration of the stride, i.e. for time β/f. Let a particular foot be on the ground from time $-\beta/2f$ to time $\beta/2f$. The vertical component F_y of the force it exerts *in this time interval* can be represented by a Fourier approximation

$$F_y = a_1 \cos (\pi ft/\beta) + b_2 \sin 2 (\pi ft/\beta) + a_3 \cos 3 (\pi ft/\beta)$$
$$+ b_4 \sin 4 (\pi ft/\beta) + \ldots . \tag{3}$$

This is a general statement which applies however F_y fluctuates. Fourier series are explained by Sneddon (1961) and in many other books. This particular Fourier series has no odd-numbered sine terms or even-numbered cosine terms because it is assumed that the force rises from zero at time $-\beta/2f$ and falls to zero at time $\beta/2f$.

Records of the vertical components of force exerted by men and other mammals in all their gaits tend to be fairly symmetrical (Fig. 4, see also Jayes & Alexander, 1978). This implies that the sine coefficients (b_2, b_4, etc.) are fairly small. The records can be imitated reasonably well by giving appro-

Fig. 5. Graphs calculated from equation 4 showing how F_y changes during the period while a foot is on the ground for different values of q. Further explanation is given in the text. From Alexander & Jayes (1978a).

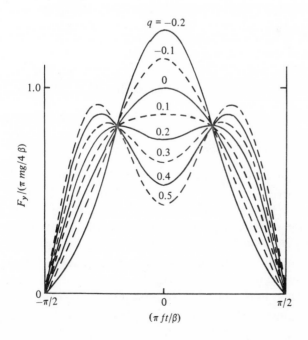

priate values to a_1 and a_3 and making all the other coefficients in equation (3) zero

$$F_y = a_1 \cos (\pi ft/\beta) + a_3 \cos 3 (\pi ft/\beta) \tag{3a}$$

If the body has mass m, gravity exerts a downward impulse mg/f on it in each stride, so each foot must exert an upward impulse $mg/2f$. Noting this and writing $q = -(a_3/a_1)$ it can be shown that

$$F_y = \frac{3\pi mg}{4\beta (3 + q)} \left[\cos \frac{\pi ft}{\beta} - q \cos \frac{3\pi ft}{\beta} \right] \tag{4}$$

Fig. 5 shows how the shape of a graph of F_y against time depends on the value of q. Negative values give bell-shaped graphs and values greater than 0.11 give graphs with two maxima. The records of F_y in Fig. 4 can be matched reasonably closely by giving q the values (a) 0.22, (b) 0.55 and (c) −0.12.

The factor q cannot be less than −0.33 or greater than 1.00, because values outside this range would make F_y negative at some stage.

Now consider the horizontal component F_x of the force exerted on the foot by the ground. It has been shown for men, dogs and sheep that this component can be represented fairly accurately by the equation

$$F_x = utF_y/k \tag{5}$$

where u is the speed of walking or running and k is a constant with the dimension of length (Alexander & Jayes, 1978b; Jayes & Alexander, 1978). For all three species and for both fore and hind feet, k lies between 1.5 and 2.2 times the height (in the standing animal) of the hip joints from the ground. It seems to be about the same for all gaits.

Equation (5), like equations (3) and (4) applies only to the period of time $-\beta/2f$ to $\beta/2f$, during which the foot is on the ground.

Optimum patterns of force

Consider a body moving up and down in simple harmonic motion. Its upward acceleration is greatest when it is at its lowest point, so the upward force on it must have a maximum then. Now look at Fig. 6a which shows the vertical components of the forces on the feet in a walk using a fairly large value of q. The total vertical force on the body is greatest when both feet are on the ground and it can be shown that the centre of mass is lowest then. The gravitational potential energy fluctuates as in Fig. 3a and the gait is a type (i) walk. In contrast, Fig. 6b represents a walk with $q = 0$. The total vertical force on the body has a minimum when both feet are on the ground, the centre of mass is highest then, and the gait is a type (iv) walk as in Fig. 3*b*.

Fig. 7 shows gaits which can be executed by a biped, moving its legs half a cycle out of phase with each other. These gaits are walks if $\beta > 0.5$ and runs if $\beta < 0.5$. All possible combinations of β and q are shown. If β and q are

Fig. 6. Schematic graphs of the vertical components of force exerted by the feet on the ground, against time, for two walking gaits. Each graph shows the forces exerted individually by the left and right feet in several steps and (by a broken line) the total of these, when both feet are on the ground. These graphs have been calculated from equation (4) using (*a*) $q = 0.4$, $\beta = 0.75$ and (*b*) $q = 0$, $\beta = 0.55$.

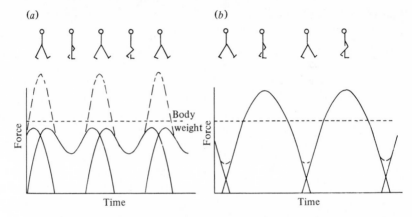

Fig. 7. (*a*) Schematic graphs of the height of the centre of mass y, against time t, for four styles of walking or running. (*b*) A graph of q against duty factor, β, showing the values which give rise to each of the styles of locomotion represented in (*a*). From Alexander & Jayes (1978a).

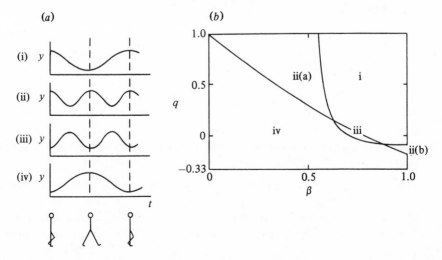

both fairly high, the gait is a type (i) walk. If q is low the gait is a type (iv) walk or a run, depending on the value of β. Zones (ii) and (iii) are occupied by gaits in which potential energy has two maxima in each half stride but the amplitude of its fluctuations is very small in zone (iii), so that the type (iii) walk is very nearly a level walk.

Alexander & Jayes (1978b) calculated the metabolic power required for bipedal walking with all possible (β, q). They used equations (2), (4) and (5) but ignored internal kinetic energy and elastic strain energy, so that their results refer to an idealised biped with light inelastic legs. They assumed that antagonistic muscles in the same leg did no work against each other but took account of work done by one leg against the other. Their method of evaluating this work was unfortunately not rigorous but the resulting errors are serious only at very high values of β which animals do not use. They showed that the power required was a function of the Froude number u^2/gk, as well as of β and q. Note that this is not the same Froude number as the one referred to on page 222: it uses the constant k of equation (5) instead of the leg length h.

Fig. 8 shows powers calculated in this way, for four different values of

Fig. 8. The metabolic power required for locomotion at various speeds. These graphs of q against β have contours showing values of $M(16\eta kf/mgu^2)$ calculated from equations (2), (4) and (5) for a biped with inelastic legs of negligible mass. They refer to the following values of u^2/gk: (a) 0.1; (b) 0.2; (c) 0.5 and (d) 1.0. From Alexander & Jayes (1978b).

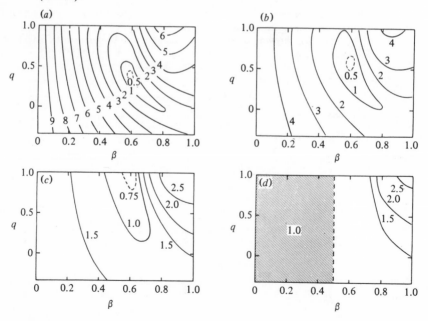

u^2/gk. When $u^2/gk < 1$ the power requirement has a minimum value which occurs at the edge of the zone of type (i) walks. The optimum value of β is always about 0.6 and the optimum value of q rises from 0.4 when $u^2/gk = 0.1$, to 1.0 when $u^2/gk = 0.5$ (Fig. 8a to c). When $u^2/gk = 1.0$ (Fig. 8d) all values of β less than 0.5 are equally good and better than values greater than 0.5. When $u^2/gk > 1$, the power requirement is least for $\beta = 0$, that is when the period of contact of the feet with the ground is infinitesimally brief. For any given finite value of β, least power is needed when q is as low as possible.

This suggests that the best technique for bipedal locomotion is to use a type (i) walk at low speeds, with $\beta \simeq 0.6$ and q increasing as speed increases. At the speed corresponding to $u^2/gk = 1$ the biped should break suddenly into a run with low values of β and q. These conclusions would probably not be altered much if account were taken of internal kinetic energy. However, if the legs are given elastic properties so that power can be saved in zone (iv) by storage of elastic strain energy, the optimum speed for breaking into a run is lower (Alexander & Jayes, 1978b).

Alexander & Jayes (1978b) did not attempt the corresponding calculations for a quadruped, but only for the mathematically simpler case of an imaginary animal with an infinite number of legs. They argued that the optimum (β, q) for a quadruped would lie between the optima for the biped and for this imaginary animal. Subsequent analysis (Alexander, 1980) shows that this argument was false. A model of quadrupedal walking shows that the optimum (β, q) for a quadruped at any given u^2/gk is approximately the same as for a biped at a slightly higher u^2/gk.

Observed walking techniques

Force-platform records show that the type (i) walk is used by men and also by the birds *Meleagris* and *Rhea* (Cavagna, Heglund & Taylor, 1977) and *Anas* (Alexander & Jayes, 1978a, reporting observations by D. J. Letten). However *Rhea* changes to a type (iv) walk at speeds above about 1.4 m s^{-1}, and Fig. 9b shows that *Coturnix* also uses both types (i) and type (iv) walks.

Fig. 9 shows values of q and β used by various animals. The points for man form two clusters, one (in zone (iv)) for running and the other for walking. Men generally walk with β between 0.55 and 0.7 and so close to the optimum values suggested by Fig. 8. They walk slowly with $q \simeq 0.2$ and faster with larger q, up to about 0.7 in very fast walking. At low speeds q is close to the values which Fig. 8 suggests are optimal for the appropriate values of u^2/gk.

At high speeds q tends to be rather less than the theoretical optimum.

The points for human walking in Fig. 9a are calculated from force records of walking at speeds of 0.9 to 2.9 m s^{-1}. All lie in zone (i), or very close to its boundary with zone (iia). Records of potential energy fluctuations confirm that the normal human walk is a type (i) walk, but some records of very high speeds (3.0–3.7 m s^{-1}) show the additional maxima characteristic of the type (ii) walk (Cavagna, Thys & Zamboni, 1976). These speeds are in the range used by athletes in walking races: men normally run at such speeds.

The points for human running in Fig. 9 have values of β between 0.3 and 0.4. The mathematical model indicates that an infinitesimal value of β would be best but this cannot be achieved in practice as it would require infinite forces on the feet. The model suggests that if β must be given a finite value q should be given a low value, and the values of q used by running men are fairly low. The speed of 2.5–3 m s^{-1} at which men generally break into a run corresponds to $u^2/gk \simeq 0.5$. The model indicates that running will not be advantageous until $u^2/gk > 1$ (Fig. 8), but the discrepancy is probably due to the model ignoring tendon elasticity (Alexander & Jayes, 1978b). Graphs of oxygen consumption against speed for men walking and running intersect at about 2.5 m s^{-1}, indicating that walking is more economical below this speed and running above it (Margaria, 1976).

Fig. 9. Graphs of q against duty factor β with points showing values obtained from force-platform records of various animals. The boundaries of the zones shown in Fig. 7b are indicated. From Alexander & Jayes (1978a).

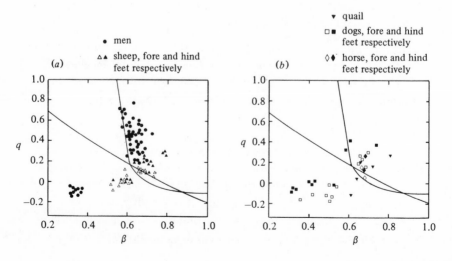

The points for dogs in Fig. 9b also form two clusters. The cluster in zone (i) includes all the points for walking and the cluster in zone (iv) refers to trotting (no points for faster gaits are shown). The change from walking to trotting occurs when u^2/gk is 0.3 or a little higher. Dogs seem not to use high values of q such as men use in fast walking.

The walking technique of sheep is still more different from that of men (Fig. 9a). The duty factor β is reduced as speed increases but there is no abrupt change of β or q (or of the phase relationship between the feet) as a walk changes to a trot.

The new model of quadrupedal locomotion (Alexander, 1980) fails to explain why dogs and sheep do not use larger values of q in fast walking.

References

Alexander, R.McN. (1976). Mechanics of bipedal locomotion. In *Perspectives in Experimental Biology 1*, ed. P. Spencer Davis, pp. 493–504. Oxford: Pergamon.

Alexander, R.McN. (1977a). Mechanics and scaling of terrestrial locomotion. In *Scale Effects in Animal Locomotion*, ed. T. J. Pedley, pp. 93–110. London: Academic Press.

Alexander, R.McN. (1977b). Terrestrial locomotion. In *Mechanics and Energetics of Animal Locomotion*, ed. R.McN. Alexander & G. Goldspink, pp. 168–203. London: Chapman & Hall.

Alexander, R.McN. (1980). Optimum walking techniques for quadrupeds and bipeds. *J. Zool. Lond.*, in the press.

Alexander, R.McN. & Bennet-Clark, H. C. (1977). Storage of elastic strain energy in muscle and other tissues. *Nature, Lond.* **265**, 114–17.

Alexander, R.McN. & Jayes, A. S. (1978a). Vertical movements in walking and running. *J. Zool. Lond.* **185**, 27–40.

Alexander, R.McN. & Jayes, A. S. (1978b). Optimum walking techniques for idealized animals. *J. Zool. Lond.* **186**, 61–81.

Alexander, R.McN., Langman, V. A. & Jayes, A. S. (1977). Fast locomotion of some African ungulates. *J. Zool. Lond.* **183**, 291–300.

Carlsöö, S. (1972). *How Man Moves*. London: Heinemann.

Cavagna, G. A. (1969). Travail mécanique dans la marche et le course. *J. Physiol., Paris*, **61** (suppl. 1), 3–42.

Cavagna, G. A., Heglund, N. C. & Taylor, C. R. (1977). Mechanical work in terrestrial locomotion: two basic mechanisms for minimizing energy expenditure. *Am. J. Physiol.* **233**, R243–R261.

Cavagna, G. A., Thys, H. & Zamboni, A. (1976). The sources of external work in level walking and running. *J. Physiol., Lond.* **262**, 639–57.

Dyson, G. H. G. (1973). *The Mechanics of Athletics*, 6th edn. London: University of London Press.

Hildebrand, M. (1976). Analysis of tetrapod gaits: general

234 R. MCN. ALEXANDER

considerations and symmetrical gaits. In *Neural Control of*
Locomotion, ed. R. M. Herman, S. Grillner, P. S. G. Stein & D. G.
Stuart, pp. 203–36. New York: Plenum.
Jayes, A. S. & Alexander, R.McN. (1978). Mechanics of locomotion of
dogs (*Canis familiaris*) and sheep (*Ovis aries*). *J. Zool. Lond.* **185**,
289–308.
McGhee, R. B. & Frank, A. A. (1968). On the stability properties of
quadruped creeping gaits. *Math. Biosci.* **3**, 331–51.
Margaria, R. (1976). *Biomechanics and Energetics of Muscular Exercise.*
Oxford: Clarendon Press.
Sneddon, I. N. (1961). *Fourier Series*. London: Routledge & Kegan
Paul.

C. R. TAYLOR

Mechanical efficiency of terrestrial locomotion: a useful concept?

Muscles as biological machines

Muscles can be thought of as 'biological machines' for generating force. They are 'tension generators' that consume energy while generating their force. If they happen to shorten while they are active, they perform mechanical work (work = force × distance). This simple principle of mechanics has enabled biologists to evaluate muscle efficiency in the same way that an engineer would evaluate the efficiency of any mechanical device that performs work or generates power (power = work or energy per unit time).

$$\% \text{ efficiency} = \frac{\text{work out}}{\text{energy in}} \times 100 = \frac{\text{power output}}{\text{power input}} \times 100$$

Vertebrate striated muscles are generally assumed to be capable of attaining efficiencies of about 25% (Hill, 1950). These values of 25% express efficiency in terms of conversion of the chemical bond energy contained in carbohydrates, fats and/or proteins into mechanical work. They have been obtained both in experiments on isolated muscles (Hill, 1964; Kushmerick, Larson & Davies, 1969) and whole animals (Margaria, 1968). For further discussion see Goldspink, this volume.

Sometimes muscle physiologists or biochemists prefer to consider the efficiency of muscles as the ratio of the mechanical work performed to the energy obtained from the breakdown of ATP. In these terms muscles seem more impressive machines, achieving efficiencies of 50 to 75% (because $\frac{1}{2}$ to $\frac{2}{3}$ of energy contained in the carbohydrates, proteins and fats is lost as heat during the formation of ATP via aerobic pathways).

Using efficiency to relate energetics and mechanics of locomotion

Attempts to relate the energetics and the mechanics of locomotion have usually made these two assumptions: (1) the major energetic cost involved in locomotion occurs when muscles shorten and perform work; and (2) that muscles perform this work at near maximal efficiency (Hill, 1950; McMahon, 1975; and Alexander, 1977). In this paper I have attempted to use some recent direct measurements of power input and power output to evaluate these two assumptions.

Does efficiency remain constant?

Measuring metabolic power input and mechanical power output

Both the metabolic power input and mechanical power output of animals have been measured as a function of speed. Metabolic power input has been calculated from measurements of steady-state oxygen consumption from a diverse assortment of vertebrates running on treadmills (Taylor, 1977). An energetic equivalent of 20.1 kJ per $1 O_2$ consumed was used in these calculations. Mechanical power output has been measured using a combination of a force platform and kinematic analysis (Cavagna & Kaneko, 1977). The force platform provides data for calculating the changes in kinetic and gravitational potential energy of an animal's centre of mass within a stride, while the kinematic analysis allows one to calculate the changes in kinetic and gravitational potential energy of the various bits and pieces of the animals relative to the centre of mass. If one assumes no elastic storage and recovery of energy, then net increases in the total mechanical energy of the system must be supplied by muscles; decreases will be lost as heat.

How does efficiency of locomotion change with running speed?

Metabolic power input increases as a linear function of speed over a wide range of speeds in most terrestrial vertebrates. This result has been well documented in man (Knuttgen, 1961; Margaria, Cerretelli, Aghemo & Sassi, 1963; Passmore & Durnin, 1955), many species of 'advanced' quadrupedal mammals (Taylor, Schmidt-Nielsen & Raab, 1970; Yousef, Robertson, Dill & Johnson, 1973), a variety of 'primitive' quadrupedal insectivores, marsupials and monotremes (Baudinette, Nagle & Scott, 1976; Crompton, Taylor & Jagger, 1978), bipedal birds (Fedak, Pinshow & Schmidt-Nielsen, 1974) and quadrupedal lizards (Bakker, 1972). There are two situations where this linear relationship has not been found: human walking (Margaria, 1938) and kangaroo hopping (Dawson & Taylor, 1973).

Mechanical power output increases as a curvilinear function of speed in terrestrial vertebrates. During the last five years Giovanni Cavagna, Norman Heglund, Michael Fedak and I have quantified the mechanical power output as a function of speed in a series of vertebrates. Fig. 1 plots data from three

Fig. 1. Mechanical power output per unit mass increases as a curvilinear function of running speed in three bipeds. Body mass has little, if any, effect on this relationship as seen by this comparison of 47-g quail, 5.7-kg turkeys and 70-kg humans. As discussed in the text, body mass does have a major effect on metabolic power input as a function of speed; the slope of the relationship between metabolic power input and speed would be more than eleven times as great in the 47-g quail as the 70-kg human. Human data are replotted from Cavagna & Kaneko (1977). Quail and turkey data are unpublished observations from a more comprehensive study of a variety of species which is still in progress by Norman Heglund, Michael Fedak, Giovanni Cavagna & C. Richard Taylor. (These data are presented as a personal communication from these investigators.)

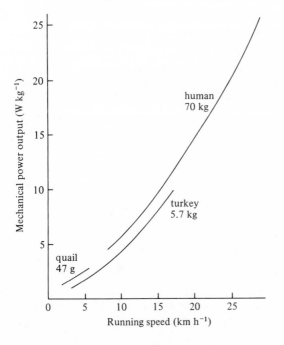

bipeds: 47-g quail, 5.7-kg turkeys (personal communication, N. C. Heglund, G. A. Cavagna, M. A. Fedak & C. R. Taylor) and 70-kg humans (Cavagna & Kaneko, 1977). In these three species (as in other species of bipeds and quadrupeds that have been studied) the total mechanical power output increases as a curvilinear function of running speed.

Efficiency of terrestrial locomotion increases with increasing speed. A simple consequence of the linear increase in metabolic power input and the curvilinear increase in power output is that mechanical efficiency of locomotion increases with speed (Fig. 2). Therefore, it does not seem possible to explain the relation between metabolic power input and speed by measuring mechanical power output and assuming a constant efficiency.

How does efficiency of locomotion change with body size?

Metabolic cost of locomotion (amount of energy expended in moving a unit mass a unit distance) varies as a regular function of body mass approximately proportional to (body mass)$^{-0.33}$ for both bipeds and quadrupeds. Our studies on energetics and mechanics of locomotion began about ten years ago with the observation that the cost of terrestrial locomotion varied as a

Fig. 2. Percentage efficiency increases both with running speed and with body mass. Percentage efficiency is defined as:

$$\frac{\text{total mechanical power output}}{\text{total metabolic power input}} \times 100$$

Total mechanical power outputs were taken from the data presented in Fig. 1, and total metabolic power inputs for humans were taken from Cavagna & Kaneko (1977) and the data for birds from Fedak *et al.* (1974).

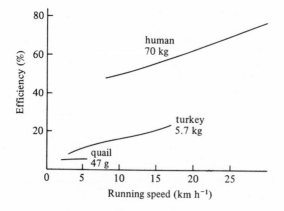

regular function of body size (Taylor *et al.*, 1970). The data base consisted of six species which happened to be around the lab at the time. The values predicted for this relationship have been close to the values obtained on additional species. Until last year the only large animals for which data existed were domestic species. Recent research in Kenya in collaboration with G. M. O. Maloiy has enabled us to remedy this problem. We were able to measure steady-state oxygen consumption as a function of speed in sixteen species of African mammals, including 260-kg eland, 100-kg wildebeest, and 100-kg waterbuck. We have recalculated the relationship between cost of locomotion and body size using this new data and other data that have been reported over the last ten years (Table 1). We now feel reasonably confident that the data base is large enough and covers a wide enough range of body size and type of mammals that it will change very little in the future as still more data becomes available. Comparing the old and new relationships by considering some 'worst-case' situations the new relationship predicts values 16% higher for a 30-g mouse running at $10 \, \mathrm{km \, h^{-1}}$ and 26% lower for a 300-kg antelope running at this speed. Since the two relationships cross, values for intermediate weight will be closer. This 'change' in predicted values seems small when viewed over a five order of magnitude range of body size. The fact that the change is small is due to both 'luck' on our first choice of animals and remarkably little variability in the relationship.

It appears that bipeds fit the same relationship as quadrupeds. New data on 100-kg ostrich and 100-kg pony collected by M. A. Fedak & H. J. Seeherman in our laboratory (personal communication) shows that both animals consume the same amount of oxygen over a wide range of speeds (from $2-25 \, \mathrm{km \, h^{-1}}$). The cost of running in the rhea, previously the largest bird from which data were available, was a little high. This skewed the relationship between cost of transport and body mass for birds so that it appeared to be higher for bipedal birds weighing over 1 kg than quadrupedal mammals of the same size (Fedak *et al.*, 1974). The absence of any difference in cost of locomotion between bipeds and quadrupeds is consistent with our observation that primates used the same amount of energy moving quadrupedally and bipedally (Taylor & Rowntree, 1973).

Mechanical work involved in moving a unit mass a unit distance appears to be independent of body mass. Fig. 1 shows that the functions relating mechanical work and running speed are similar for 47-g quail, 5.7-kg turkeys and 70-kg humans. We are still expanding our data base. However, we now feel reasonably confident that this situation will be true in general terms for most bipeds and quadrupeds.

Table 1. Old and new relationship between cost of terrestrial locomotion and body size (units are $\dot{V}O_2$ in ml $O_2\,g^{-1}h^{-1}$; M_b is body mass in g, V is running speed in km h^{-1})

	Predicted $\dot{V}O_2$ for 30-g animal running at 10 km h^{-1} (ml O_2 $g^{-1}h^{-1}$)	Predicted $\dot{V}O_2$ for 300-kg animal running at 10 km h^{-1}
Old relationship (Taylor et al., 1970) $\dot{V}O_2 = 8.46\,M_b^{-0.4}V + 6\,M_b^{-0.25}$	24.3	0.80
New relationship (C. R. Taylor, G. M. O. Maloiy, H. J. Seeherman, N. C. Heglund, V. A. Langman and J. Kamau, personal communication) $\dot{V}O_2 = 4.22\,M_b^{-0.30}V + 13.7\,M_b^{-0.39}$	18.8	1.06

Efficiency of terrestrial locomotion increases with increasing size. This is apparent from the data presented in Fig. 2 from 47-g quail, 5.7-kg turkeys and 70-kg humans. Efficiency of locomotion has increased by more than ten-fold with this three orders of magnitude increase in body size.

How can energetics and mechanics of locomotion be related?

Pendulums and springs

Since the efficiency of locomotion is not constant, and varies both with speed of locomotion and body size, one must consider what the muscles of running animals are doing in order to understand the energetics of locomotion. Cavagna, Heglund & Taylor (1977) have shown that an assortment of vertebrates, including man, use two basic mechanisms for minimising energy expenditure during terrestrial locomotion. Walking involves an alternate transfer between gravitational potential energy and kinetic energy within each stride, such as takes place in a pendulum. This transfer is greatest at intermediate walking speeds and can account for up to 70% of the total energy changes taking place within a stride, leaving only 30% to be supplied by muscles. In running, hopping and trotting no kinetic–gravitational energy transfer takes place. Energy is conserved by another mechanism: an elastic 'bounce' of the body. Galloping animals utilise a combination of these two energy-conserving mechanisms. A fuller discussion of this aspect is given in Alexander, this volume.

What do muscles do during locomotion?

Much of the energy used by muscles during normal locomotion may not be used for performing positive mechanical work. For example, significant amounts of metabolic energy may be required for active muscles to exert force without changing length in order to stabilise the joints. Also, in order for energy to be stored elastically in muscles and their tendons, the muscles must be active while they are being stretched. This type of activity also requires energy. For further consideration see Goldspink, this volume.

We have measured the electrical activity of fourteen major locomotory muscles as a function of gait and speed in dogs. We find that many of the muscles that are involved in joint stabilisation during a walk show either very small length changes or none at all. Also, as the dogs increase speed during a walk or trot, propulsive muscles such as the *biceps femoris* become active while the limb is still moving forward (i.e. during the recovery stroke when the muscle is being stretched). This could allow some of the kinetic energy of the limb to be stored elastically as the limb decelerates. Presumably some of this energy can be recovered as the limb reverses direction. Activity while the muscle is being stretched becomes increasingly important with increasing speed. Muscles such as the *triceps brachii* show a similar pattern. They also become active while being stretched, allowing energy to be stored elastically as the dog falls and decelerates on landing, and recovered as the dog lifts and reaccelerates during the subsequent part of the stride.

Looking for an explanation of the energetics of terrestrial locomotion

Although the efficiency of active muscles that shorten and perform work may be the same in vertebrates of different sizes and in muscles with different contraction velocities, one might anticipate that the cost of maintaining tension and of 'alternate stretching and shortening' would vary with body size. The intrinsic velocity of muscles (maximum contraction velocity measured in lengths per second of muscle under zero load) is the one property of vertebrate striated muscle which varies markedly with body size and from muscle to muscle. Intrinsic velocity can differ by more than 700-fold among vertebrate striated muscle (Prosser, 1973). Mass-specific power output, cost of maintaining tension and mysoin ATPase activity seem to be proportional to intrinsic velocity (Hill, 1964; Awan & Goldspink, 1970; Close, 1972; and M. Crowe, H. J. Seeherman & M. J. Kushmerick, personal communication). If this is the case then cost of maintaining tension and operating the pendulum and spring system is going to increase dramatically as intrinsic velocity increases with decreasing body size (since Ruegg (1971) and Close (1972) report mass-specific myosin ATPase activity is proportional to body mass $^{-0.125}$). Thus smaller animals should have to expend much more energy in operating their 'pendulums and springs' than large animals. Heglund is collaborating with Kushmerick and Cavagna in an attempt to measure the energetics and mechanics of these three types of muscular activity (when muscles shorten, lengthen, and stay the same length). Hopefully this may provide us with a link between energetics and mechanics of muscles which will help to explain the energetics of locomotion in mechanical terms.

References

Alexander, R. McN. (1977). Terrestrial locomotion. In *Mechanics and Energetics of Animal Locomotion*, ed. R. McN. Alexander & G. Goldspink, pp. 168–203. London: Chapman & Hall.

Awan, M. Z. & Goldspink, G. (1970). Energy utilization by mammalian fast and slow muscle in doing external work. *Biochem. Biophys. Acta.* **216**, 229–30.

Bakker, R. T. (1972). Locomotory energetics of lizards and mammals compared. *The Physiologist*, **15**, 15.

Baudinette, R. V., Nagle, K. A. & Scott, R. A. D. (1976). Locomotory energetics in Dasyurid marsupials. *J. comp. Physiol.* **109**, 158–68.

Cavagna, G. A., Heglund, N. C. & Taylor, C. R. (1977). Mechanical work in terrestrial locomotion: two basic mechanisms for minimizing energy expenditure. *Am. J. Physiol.* **233**, R243–61.

Cavagna, G. A. & Kaneko, M. (1977). Mechanical work and efficiency in level walking and running. *J. Physiol., Lond.* **268**, 467–81.

Close, R. I. (1972). Dynamic properties of mammalian skeletal muscles. *Physiol. Rev.* **52**, 129–97.

Crompton, A. W., Taylor, C. R. & Jagger, J. A. (1978). Evolution of homeothermy in mammals. *Nature, Lond.* **242**, 333–6.

Dawson, T. J. & Taylor, C. R. (1973). Energetic cost of locomotion in kangaroos. *Nature, Lond.* **246**, 313–14.

Fedak, M. A., Pinshow, B. & Schmidt-Nielsen, K. (1974). Energy cost of bipedal running. *Am. J. Physiol.* **227**, 1038–44.

Hill, A. V. (1950). The dimensions of animals and their muscular dynamics. *Science Prog.* **38**, 209.

Hill, A. V. (1964). The efficiency of mechanical power development during muscular shortening. *Proc. R. Soc., Ser. B,* **159**, 319–24.

Knuttgen, H. G. (1961). Oxygen uptake and pulse rate while running with undetermined and determined stride lengths at different speeds. *Acta Physiol. Scand.* **52**, 366–71.

Kushmerick, M. J., Larson, R. E. & Davies, R. E. (1969). The chemical energies of muscle contraction. 1. Activation heat, heat of shortening and ATP utilization of activation–relaxation process. *Proc. R. Soc., Ser. B,* **174**, 293–314.

Margaria, R. (1938). Sulla fisiologia e specialmente sul consumo energetico della marcia e della corsa a varie velocita ed inclinazioni del terreno. *Atti Accad. Naz. Lincei Mem. Classe Sci. Fis. Mat. Nat. Sez.* (III) **7**, 299–368.

Margaria, R. (1968). Positive and negative work performances and their efficiencies in human locomotion. *Int. Z. angew. Physiol.* **25**, 339–51.

Margaria, R., Cerretelli, P., Aghemo, P. & Sassi, G. (1963). Energy cost of running. *J. appl. Physiol.* **18**, 367–70.

McMahon, T. A. (1975). Using body size to understand the structural design of animals: quadrupedal locomotion. *J. appl. Physiol.* **39**, 619–27.

Passmore, R. & Durnin, J. V. G. A. (1955). Human energy expenditure. *Physiol. Rev.* **35**, 801–40.

Prosser, C. L. (ed.) (1973). *Comparative Animal Physiology.* Philadelphia: W. B. Saunders.

Ruegg, J. C. (1971). Smooth muscle tone. *Physiol. Rev.* **51**, 201–48.

Taylor, C. R. (1977). The energetics of terrestrial locomotion and body size in vertebrates. In *Scale Effects in Animal Locomotion*, ed. T. J. Pedley, pp. 127–41. London: Academic Press.

Taylor, C. R. & Rowntree, V. J. (1973). Running on two or four legs: which consumes more energy? *Science, NY,* **179**, 186–7.

Taylor, C. R., Schmidt-Nielsen, K. & Raab, J. L. (1970). Scaling of energy cost of running to body size in mammals. *Am. J. Physiol.* **219**, 1104–7.

Yousef, M. K., Robertson, W. D., Dill, D. B. & Johnson, H. J. (1973). Energetic cost of running in the antelope ground squirrel *Ammospermophilus leucurus. Physiol. Zool.* **46**, 139–47.

INDEX